A SHEARWATER BOOK

Revolution on the Range

REVOLUTION

ISLANDPRESS / Shearwater Books

Washington • *Covelo* • *London*

ON THE RANGE

The Rise of a New Ranch
in the American West

by Courtney White

A Shearwater Book
Published by Island Press

Copyright © 2008 Courtney White

All rights reserved under International and Pan-American
Copyright Conventions. No part of this book may be reproduced
in any form or by any means without permission in writing from
the publisher: Island Press, 1718 Connecticut Ave., NW, Suite 300
Washington, DC 20009.

SHEARWATER BOOKS is a trademark of The Center for Resource Economics.

Library of Congress Cataloging-in-Publication data.
White, Joseph Courtney, 1960–
Revolution on the range : the rise of a new ranch in the American West /
by Courtney White.
p. cm.
"Shearwater Books."
Includes bibliographical references and index.
ISBN-13: 978-1-59726-174-6 (cloth : alk. paper)
ISBN-10: 1-59726-174-2 (cloth : alk. paper)
1. Ranching—West (U.S.) 2. Grassland restoration—West (U.S.)
I. Title.
SF196.U5W46 2008
636'.010978—dc22   2007046946

British Cataloguing-in-Publication data available.

Printed on recycled, acid-free paper ✤

Design by David Bullen

Manufactured in the United States of America

10 9 8 7 6 5 4 3 2 1

Keywords: cattle, beef, grazing, Rocky Mountain West, cowboys,
environment, conservation, federal land, private land, resource
management, restoration, Radical Center, Quivira Coalition

To Gen,

who was there, side by side, every step of the way.

CONTENTS

PART THREE: **THE BIG PICTURE**

PROLOGUE

Out beyond the ideas of rightdoing and wrongdoing
is a field. I'll meet you there.

RUMI

In 1996, I had an anguished question on my mind: why didn't environmentalists and ranchers get along better? In theory they shared many of the same hopes and fears—a love of wildlife, a deep respect for nature, an appreciation for a life lived outdoors, and a common concern for healthy water, food, fiber, and liberty.

That was the theory anyway. The reality was that by the early 1990s environmentalists and ranchers, along with loggers, federal land managers, elected officials, private citizens, and others in the American West, were locked in a bitter struggle with one another, exemplified by two popular bumper stickers of the era: "Cattle-free by '93!" shouted one. "Cattle galore by '94!" retorted the other.

I felt anguished because this fight had all the hallmarks of a tragedy: both sides, and all of us in between, seemed destined to lose what was most valued by everyone—the health and diversity of the West's wide open spaces. And it wasn't just the West: the hardheadedness of this particular fight reflected other divides in the nation at the time—the "red" and "blue" split, for instance, that would soon engulf our national politics.

The causes of the conflict between ranchers and environmentalists were more social and historical than ecological, in my opinion. Certainly, overgrazing by livestock in the arid West had damaged, and in some cases irreparably altered, native plant and animal communities, raising legitimate cries of alarm. However, other issues fueled the grazing debate to a larger extent, including class, political power, and prejudice. Ignorance played a role too, unfortunately—a point brought home in force one day when an environmental activist told me, with a straight face, that cattle were "immoral animals."

The struggle focused primarily on the publicly owned half of the American West's one million square miles, including the national forests, rangelands, and wildlife refuges. The fundamental issue was influence. For a century or more, these federal lands were in the de facto control of those who lived near them and worked on them—ranchers, principally—and who operated largely without oversight. After World War II, however, influence began to shift to a new breed of westerner—hikers, fishermen, day-trippers, and other types of often urban-based recreationalists. At first, their influence was largely economic, but over time it grew politically, especially as the populations of western cities boomed.

Concurrently, a concern for the welfare of nature in the form of a resurgent conservation movement—now called environmentalism—started to blossom across the nation. Increasingly, the attention of activists turned toward actual and perceived abuses of the public domain, including clear-cut logging, open-pit mining, and overgrazing. The alarms they raised contributed to a raft of consequential environmental legislation passed by Congress and signed into law by President Richard Nixon, including the National Environmental Policy Act, the Endangered Species Act, and an early version of the Clean Water Act, as well as a bill creating the Environmental Protection Agency.

The downside, however, of all this activism and bill-passing was the commencement of a kind of tribal warfare between denizens of the "Old" West and advocates of the "New," with lassos on one side, and lattes on the other. Caught in the middle were the employees of the federal land management agencies—the Forest Service (national forests), the Bureau of Land Management (rangelands), and the Fish & Wildlife Service (refuges). The "feds," once considered by environmentalists to be in the pocket of ranching, mining, and timber interests, by the 1980s were viewed by ranchers, miners, and loggers as allies of the environmentalists instead. This meant that federal employees found themselves in the crosshairs of both sides.

Meanwhile, across the West, accelerating suburban and exurban (ranchette) growth shared the same source: former farm and ranch land. When making their case against cattle, environmental activists frequently pointed out that half of the West is publicly owned, and therefore should be managed with public goals in mind. But they overlooked the flip side of their own statistic—the other half of the West is privately owned, much of it by ranchers. Deliberate or not, by weakening ranchers, environmentalists abetted the very thing they decried loudest about the New West—its breakup by sprawl and other forms of land fragmentation.

There were other reasons to worry about the fate of ranchers besides the loss of open space. Healthy food, for one thing. As writer and farmer Wendell Berry has repeatedly observed, eating is an agricultural act. We all do it at least three times a day, which is why it's worth thinking long and hard about where our food comes from, who grows it, under what conditions it is produced, and what the consequences are of letting a global, industrialized food system fill our bellies. The family rancher, by contrast, could, I knew, produce healthy, locally grown food under humane conditions at a reasonable price. Throw in good stewardship of the land and you have the possibility of an unbeatable combination, which is why the prospect of eliminating the family rancher, even on public land, was so distressing.

Ranchers also had legitimate historical and cultural claims to existence. In northern New Mexico, where I live, the ranching tradition stretches back 400 years—and much farther if you trace it back to Spain. Any knowledgeable historian or anthropologist would agree that ranching is an important subset of American society—and not because of its influence on Hollywood, Nashville, or Madison Avenue. Ranchers have been a critical part of America's ethnic and historical tapestry, and remain so to this day.

Lastly, ranching mattered, I recognized, because work matters and because land matters. Although I had spent a lot of time backpacking as a youth, enjoying the recreational fruits of our robust economy, I also spent many summers surveying the desert of southern Arizona as a professional archaeologist. It was a form of hiking, but it was also *work*—and as a consequence I came to appreciate the value of labor on the land. I gained a physical and emotional relationship to nature that wasn't play-based, and this made a huge difference.

For all these reasons, the conflict between ranchers and environmentalists began to look like a tragedy of rather serious proportions to me.

By the mid-1990s, in fact, the feud between industry and activists had reached a dispiriting crescendo. Newspaper headlines reported a seemingly endless cycle of unhappy news: effigies of environmentalists hanging from street lamps; road building equipment disabled in the dead of night; federal property attacked by anonymous assailants; hiking trails booby-trapped with explosives; trees "spiked" with large nails to prevent their harvest; cattle shot; endangered species threatened by a campaign of "shoot, shovel, and shut up"; public meetings dissolving into shouting matches; shadowy militias organizing in remote locations; federal raids ending disastrously; livelihoods ruined by lawsuits; and so on.

Emblematic of the times was a lengthy brawl in the mountains above Silver City, in southwestern New Mexico. Called the Diamond Bar fight—for the 145,000-acre Forest Service allotment (ranch) on which the fight took place—it featured an angry young ranching couple, Kit and Sherry Laney, who were determined to prevail over the U.S. Forest Service, and an even angrier local environmentalist equally determined to put them out of business. Public lands are divided into allotments of varying sizes, which are generally attached to a base (private) property owned by the rancher. A grazing permit is issued by the federal agency for that allotment and contains conditions, including allowable numbers of cattle, by which the livestock operation must abide. On the surface, the fight focused on the government's attempt to force the Laneys to abide by certain regulations, including a recent reduction in the amount of cattle they could run on the allotment. These were restrictions that the young ranchers rejected and that environmentalists demanded be upheld. The real issue, however, was power: who would win and who would lose.

Stuck in the middle was a fumbling federal bureaucracy whose attempts at compromise succeeded only in stoking the conflict. Charges, countercharges, lawsuits, appeals, and threats flew in all directions as both sides marshaled their supporters for what appeared to be the Final Showdown over livestock grazing on public land in the Southwest.

In the end, the Laneys lost. Acting unwisely on poor legal advice, they refused to sign their grazing permit, asserting that the government had no right to regulate them, which meant they were breaking the law. When a judge upheld the Forest Service's position, the Laneys lost their permit and their ranch, as well as their livelihood.

Environmentalists were elated. A significant corner, they said, had been turned in the struggle over public lands in the West.

To this particular environmentalist, however, there was no cheer in the court's verdict. I did not join the celebrations when the victorious activists came to Santa Fe, but neither did I mourn the demise of the young ranchers, who had arrogantly thumbed their noses at public opinion. Instead, I just felt depressed. There were no winners in the Diamond Bar fight, only losers, including all the spectators. That's because nothing had been gained—lives had been ruined, not enriched; land had been abandoned, instead of stewarded properly; bad blood had been created, instead of hope; anger ruled, not joy.

My anguished question involved more than just bad blood between ranchers and environmentalists, however. The Diamond Bar fight fit a national mood in the mid-1990s that had suddenly veered onto the rocky shoals of partisanship, confrontation, and political brinkmanship. From the jeremiads of talk-radio hosts, which capitalized on the new rancor emanating from Washington, D.C., to repeated shutdowns of the federal government, America seemed suddenly caught in a destructive tug-of-war between Wrongdoing (them) and Rightdoing (us), with no room for anybody in between.

And the more we yelled at one another, the deeper my spirits sank. Then one day something snapped inside me and I knew I had to act.

It happened on April 19, 1995—the day Timothy McVeigh blew up the Murrah Federal building in downtown Oklahoma City, killing 168 innocent people, including 19 children, and injuring more than 800 people. I worked for the National Park Service as an archaeologist at the time, as did my wife, and I remember vividly my reaction as I listened, stunned, to the news report of the bombing coming in over a radio in the office.

At first, I was mortified, and then I grew angry, but not just at McVeigh. I was angry at the whole culture of conflict and odium represented by this horrible tragedy. McVeigh wasn't simply a madman—he had *motivation*, as he explained later. He *hated*. It didn't matter that the object of his ire was the federal government, what mattered was the emotion itself—the same negativity circulating around the nation at the time; the same emotion at work in the mountains above Silver City. Although some pundits later denied any causal connection between McVeigh's act of terrorism and the partisan cultural climate in America, I knew the bombing had happened for a reason.

It happened because it was OK to hate.

I had to do something, but what? The previous fall, alarmed by the "Republican Revolution" in the 1994 midterm elections and the declared intention of its leaders to roll back twenty-five years of critical environmental legislation, I had called a representative of the Sierra Club to volunteer my services. I was quickly recruited as a foot soldier for the Club's local group in Santa Fe and less than two months later I was sent into battle at the state capitol during the legislative session, assigned the job of fighting "takings" legislation—a complex legalistic assault on the public good by private property rights advocates. For my efforts, and to my surprise, I wound up on a stage in an auditorium that summer debating takings with the executive director of the New Mexico Cattlegrowers' Association in front of a large crowd of businesspeople.

I have no idea who won the debate, though I recall being embarrassed at my decision to wear cowboy boots. It was an attempt at an ironic statement, but it came across as just plain silly. I also recall the empty feeling the debate left inside of me. Intellectually, I understood the need to push back against wrongdoers, as the environmental movement was successfully doing against the Republican agenda in Washington at the time, but emotionally I felt adrift.

Eventually, an unexpected opportunity to act on my anguish came. Walking into a statewide meeting of the Sierra Club one day, held in the former mining boomtown of Kingston, New Mexico (and not far from the Diamond Bar allotment), I saw a cowboy hat sitting on a table. It belonged, I learned, to Jim Winder, who lived and ranched nearby. If that wasn't surprise enough, I was told Jim was there because he had accepted the invitation of the chair, Gwen Wardwell, to become a member of the Executive Committee.

A rancher on the statewide Executive Committee of the Sierra Club? And a Republican to boot! What was going on here?

Jim boasted that he ranched in a new, ecologically friendly style. He bunched his cattle together into one herd, he said, and kept them on the move so that any particular patch of ground would be grazed only once a year, mimicking the manner in which bison covered the land. He didn't kill coyotes. In fact, he didn't even mind wolves, because bunched-up cows can protect themselves. There was more: because he ranched for rangeland health, Jim said, he got along great with government employees, he had more water in his streams, and most importantly, he was making money.

It sounded too good to be true.

Curious about this newfangled ranching, in early 1996 I joined a tour of the Winder family ranch Jim had organized for his fellow Sierra Clubbers. Attending as well was an antigrazing activist named Tony Merten, who had recently transplanted himself from Colorado to a remote part of southern New Mexico. I didn't know it at the time, but Tony was the prime suspect in a spate of cattle murders in the area. It would be an investigation with tragic consequences. Whether from fear of a potential indictment, mental instability, or a deep sense of despair for the fate of the planet (or all three), Tony would commit suicide a little more than a month after the tour of Jim's ranch.

On that day, however, it quickly became clear to me that Tony's mission was to provoke Jim into a confrontation. He obnoxiously challenged nearly every positive statement Jim made, whether it was about cattle, grass, or termites (a favorite subject of Jim's). It didn't work. Jim parried each attack with a patient explanation of ecological principles and a fine sense of humor. In fact, it was obvious that Jim knew far more about the environment than any environmentalist on the tour, myself especially. He was far funnier too.

Impressed, embarrassed, and perplexed, upon my return home I picked up *Beyond the Rangeland Conflict: Toward a West That Works*, a book by environmental activist Dan Dagget. In it, I learned that there were other ranchers of Jim's stripe across the West—people managing for healthy ecosystems through progressive cattle management and collaboration. The book confirmed what I saw on Jim's ranch: thick grass, healthy riparian areas, young plants, wildlife, open space—all the things I *said* I wanted as a conservationist. Of course, I saw livestock too.

The anguished question began to grow.

Inspired as much by his performance as by his knowledge, I called Jim up and asked him if we should try to create a neutral forum where anyone who loved the land, wildlife, and cultures of the Southwest could meet, look, learn, and listen. He enthusiastically endorsed the idea. We were joined by Barbara Johnson, another Sierra Club activist. The three of us quickly decided that there was no point in engaging the extremes on either side of the grazing debate. Instead, we would walk to a new field, beyond the continuum of argument, where we would wave our arms and ask people to join us. Jim called this place the "third position."

I called it the New Ranch.

I wrote a definition: "The New Ranch describes an emerging progressive ranching movement that operates on the principle that the natural processes that sustain wildlife habitat, biological diversity and functioning watersheds are the same processes that make land productive for livestock. New Ranches are ranches where grasslands are productive and diverse, where erosion has diminished, where streams and springs, once dry, now flow, where wildlife is more abundant, and where landowners are more profitable as a result."

The New Ranch became the foundation for an exploration of our larger goal: "to explore our common interests instead of argue our differences," in the words of Bill deBuys, a conservationist and leader in the collaborative movement in New Mexico.

Exploring common interests was an idea gaining traction at the time. In pockets across the West, groups of ranchers, federal managers, and environmentalists had been attempting to start meaningful dialogues. One highly successful effort was located in the "bootheel" of southwestern New Mexico, where a diverse group had come together to put ecologically beneficial fire back on the land as well as to shield private lands from the predatory attention of subdividers. They called themselves the Malpai Borderlands Group.

We called ourselves The Quivira Coalition. On Spanish colonial maps of the Southwest from the 1600s, "Quivira" designated unexplored territory.

Following the lead of other "common ground" efforts, we vowed to avoid lawsuits and legislation, sticking instead to the grassroots— literally the "grass" and the "roots." It was our belief that the grazing debate needed to start over at the place it mattered most—on the ground. We knew it was a gamble. When we organized our first workshop in a church in Santa Fe in June 1997, we sent out notices to every moderate rancher, environmentalist, land manager, and scientist we knew in New Mexico. Then we crossed our fingers. When fifty people showed up, we knew we weren't going to be alone in our little field.

In the years that followed, as the grazing debate faded in the region and as hope and trust began to grow alongside the wildflowers and bunchgrasses, an answer to my anguished question began to reveal itself. Ranchers and environmentalists *could* get along, and in places *did*, especially where the dialogue started with soil, grass, and water. Peace, in other words, was possible; and as a result, progress was possible as well.

But there was more. In fact, a new anguished question had begun to grow.

It started with a map I saw of a 500,000-acre watershed in southern Arizona. It was a map of rangeland health, meaning it viewed the land from a functional perspective—from the angle of soil, grass, and water. According to the analysis represented on the map, significant amounts of the watershed were in poor condition, including big portions of a national wildlife refuge, which had not been grazed by cattle in sixteen years. "Goodness," I thought to myself after studying it, "how much of the rest of the West is in this condition?"

This issue hit home one day as I walked up a deep arroyo (wash) on a ranch in western New Mexico. As I came to the boundary between the private land and the Forest Service property, I saw a barbed wire fence, complete with fence posts, suspended ten feet above my head, stretching across the arroyo. I knew from a conversation with the rancher that the fence was built in 1935—and the posts rested on the ground. In less than seventy years, in other words, the system had unraveled—washed away.

Poor grazing management played a role, undoubtedly. When the ground lacks a vigorous cover of healthy vegetation, its exposure to the erosive effects of pounding rain and rushing water dramatically increases. But my work with Jim Winder had taught me that cattle could be managed in a positive manner for the health of land. Jim—and others—taught me that cows weren't the problem, poor management was. Things could be different.

Looking up at the fence suspended above my head that day, I began to ask questions: How do we restore this land to health? What are the tools? How do we pay for it?

Fortunately, a pattern of answers was already visible. The work of the New Ranchers demonstrated that sustainable and regenerative land management was not only possible, it could be profitable too. At the same time, new restoration methods had been developed, which also worked within "nature's model" of land health, providing relatively simple and cost-effective strategies for reversing ecosystem decline.

In short, peace making led me to see how healthy land and healthy relationships could be restored, one acre at a time.

The chapters in this book—representing a personal journey—are my attempt to illustrate how ranching and environmentalism are changing in the West, and with them, the West itself—and with the West, the nation too, possibly. The people profiled not only ask questions of their own, they also form part of a pattern of solutions. Linked together, they are part of an intriguing mosaic of human creativity, energy, and hopefulness.

This is a book about relationships—among people, between people and land, among ecological processes—and their resilience. When I first started writing the essays that eventually led to this book, I wanted to do nothing more than hold up what I considered to be my most valuable discoveries. Over time, however, I realized that the discoveries were not nearly as important as the relationships that lay behind them. I came to see that, whether in the American West or beyond, healthy *things*—cattle, wolves, watersheds, communities, economies, nations—depend on a foundation of healthy relationships. And often the key to enhancing the resilience of those relationships is to create a field beyond rightdoing and wrongdoing.

I'll see you there.

Revolution on the Range

PART ONE

The New Ranch

Chapter One

THE NEW RANCH

Ranching is one of the few western occupations
that have been renewable and have produced a
continuing way of life.

WALLACE STEGNER

It was a bad year to be a blade of grass.

In 2002, the winter snows were late and meager, part of an emerging period of drought, experts said. Then May and June exploded into flame. Catastrophic "crown" fires scorched more than a million acres of evergreens in the Four Corners states—New Mexico, Arizona, Colorado, and Utah—making it a bad year to be a tree too.

The monsoon rains then failed to arrive in July, and by mid-August hope for a "green-up" had vanished. The land looked tired, shriveled, and beat-up. It was hard to tell which plants were alive, dormant or stunned, and which were dead. One range professional speculated that perhaps as many as sixty percent of the native bunchgrasses in New Mexico would die. It was bad news for the ranchers he knew and cared about, insult added to injury in an industry already beset by one seemingly intractable challenge after another.

For some, it was the final blow. Ranching in the American West, much like the grass on which it depended that year, has been struggling for survival. Persistently poor economics, tenacious opponents, shifting values in public-land use, changing demographics, decreased political influence, and the temptation of rapidly rising private land values have all combined to push ranching right to the edge. And not just ranching; according to one analysis the number of natural resource jobs, including agriculture, as a *share* of total employment in the Rocky Mountain West has declined by two-thirds since the mid-1970s. Today, less than one in thirty jobs in the region is in logging, mining, or agriculture. This fits a national trend. By 1993, the U.S. Census Bureau had dropped its long-standing survey of farm residents. The farm population across the nation had dwindled from 40 percent of households in 1900 to a statistically insignificant 2 percent by 1990. The Bureau decided that a survey was no longer relevant.

If the experts are correct—that the current multiyear drought could rival the decade-long "megadrought" of the 1950s for ecological, and thus economic, devastation—the tenuous grip of ranchers on the future will be loosened further, perhaps permanently. The ubiquitous "last cowboys," mythologized in a seemingly endless stream of tabletop photography books, could ride into their final sunset once and for all.

Or would they?

After all, for millions of years grass has always managed to return and flourish. James Ingalls, U.S. senator from Kansas (1873–1891) once wrote:

> Grass is the forgiveness of nature—her constant benediction. Fields trampled with battle, saturated with blood, torn with the ruts of cannon, grow green again with grass, and carnage is forgotten. Streets abandoned by traffic become grass grown like rural lanes, and are obliterated; forests decay, harvests perish, flowers vanish, but grass is immortal.

Few understand these words better than ranchers, who, because their cattle require grass, depend on the forgiveness of nature for a livelihood while simultaneously nurturing its beneficence. And like grass, ranching's adaptive response to adversity over the years has been patience—to outlast its troubles. The key to survival for both has been endurance—the ability to hold things together until the next rainstorm. Evolution favors grit.

Or at least it used to.

Today, grit may still rule for grass, but for ranchers it has become more hindrance than help. "Ranching selects for stubbornness," a friend of mine likes to say. While admiring ranching and ranchers, he does not intend his quip to be taken as a tribute. What he means is this: stubbornness is not adaptive when it means rejecting new ideas or not adjusting to evolving values in a rapidly changing world.

This is where ranching and grass ultimately part ways: unlike grass, ranching may not be immortal.

Fortunately, a growing number of ranchers understand this and are embracing a cluster of new ideas and methods, often with the happy result of increased profits, restored land health, and repaired relationships with others. I call their work the New Ranch. But what did it mean exactly? What were the "new" things ranchers were doing to stay in business while neighboring enterprises went under? How did they differ from "new" ranch to "new" ranch? What were the commonalities? What was the key? Technology, ideas, economics, increased attention to ecology, or all of the above?

During that summer of fire and heat I decided to take a 1,400-mile drive from Santa Fe to Lander, Wyoming, and back, visiting four ranch families in order to see the New Ranch up close. Initially, I wanted to know if ranching would survive this latest turn of the evolutionary wheel. Was it still renewable, as Stegner once observed, or is the New West destined to redefine a ranch as a mobile home park and a subdivision? But I also wanted to discover the outline of the future, and, with a little luck, find my real objective—hope—which, like grass, is sometimes required to lie quietly, waiting for rain.

The James Ranch *North of Durango, Colorado*

One of the first things you notice about the James Ranch is how busy the water is. Everywhere you turn, there is water flowing, filling, spilling, irrigating, laughing. Whether it is the big, fast-flowing community ditch, the noisy network of smaller irrigation ditches, the deliberate spill of water on pasture, the refreshing fish ponds, or the low roar of the muscular Animas River, take a walk in any direction on the ranch during the summer and you are destined to intercept water at work. It is purposeful water too, growing trees, cooling chickens, quenching cattle, raising vegetables, and, above all, sustaining grass.

All this energy on one ranch is no coincidence—busy water is a good metaphor for the James family. The purposefulness starts at the top. Tall, handsome, and quick to smile, David James grew up in southern California, where his father lived the American dream as a successful engineer and inventor, dabbling a bit in ranching and agriculture on the side. David attended the University of Redlands in the late 1950s, where he majored in business, but cattle got into his blood, and he spent every summer on a ranch. David met Kay, a city girl, at Redlands, and after getting hitched, they decided to pursue their dream: raise a large family in a rural setting.

In 1961, they bought a small ranch on the Animas River, twelve miles north of the sleepy town of Durango, located in a picturesque valley in mountainous southwestern Colorado, and got busy raising five children and hundreds of cows. Durango was in transition at the time from a mining and agricultural center to what it is today: a mecca for tourists, environmentalists, outdoor enthusiasts, students, retirees, and real estate brokers. Land along the river was productive for cattle and still relatively cheap in 1961, though a new crop—subdivisions—would be planted soon enough.

Not long after arriving, David secured a permit from the U.S. Forest Service to graze cattle on the nearby national forest. The permit allowed him to run a certain number of cattle on a forest allotment. Once on the forest, he managed his animals in the manner he had been taught: uncontrolled, continuous grazing.

"In the beginning, I ranched like everyone else," said David, referring to his management style, "which means I lost money."

David followed what is sometimes called the "Columbus school" of ranching: turn the cows out in May, and go discover them in October. It's a strategy that often leads to overgrazing, especially along creeks and rivers, where cattle like to linger. Plants, once bitten, need time to recover and grow before being bitten again. If they are bitten too frequently, especially in dry times, they can use up their root reserves and die, which is bad news for the cattle (not to mention the plant). Because ranchers often work on a razor-thin profit margin, it doesn't take too many months of drought and overgrazing before the bottom line begins to wither too.

Grass may be patient, but bankers are not.

Through the 1970s, David's ranchlands, and his business, were on a downward spiral.

When the Forest Service cut back his allowable cattle numbers, as they invariably did in years of drought, the only option available to David was to run them on the home ranch, which meant running the risk of overgrazing his seven hundred acres of private land. Meanwhile, the costs of operating the ranch kept rising. It was a no-win bind typical of many ranches in the West.

"I thought the answer was to work harder," he recalled, "but that was exactly the wrong thing to do."

Slowly, David came to realize that he was depleting the land, and himself, to the point of no return. By 1978 things became so desperate that the family was forced to develop a sizeable portion of their property, visible from the highway today, as a residential subdivision called, ironically, "The Ranch." It was a painful moment in their lives.

"I never wanted to do that again," said David, "so I began to look for another way."

In 1990, David enrolled in a seminar taught by Kirk Gadzia, a certified instructor in what was then called Holistic Resource Management—a method of cattle management that emphasizes tight control over the timing, intensity, and frequency of cattle impact on the land, mimicking the behavior of wild herbivores, such as bison, so that both the land and the animals remain healthy. "Timing" refers not only to the time of year but how much time, measured in days rather than the standard unit of months, the cattle spend in a particular paddock. "Intensity" means how many animals are in the herd for that period of time. "Frequency" means how long the land is rested before a herd returns.

All three elements are carefully mapped out on a chart, which is why this strategy of ranching is often called "planned grazing." The movement of the cattle herd from one paddock or pasture to another is carefully designed, often with the seasonal needs of wildlife in mind. Paddocks can range from a few acres to hundreds of acres in size, depending on many variables, and are often created with two-strand, solar-powered electric fencing, which is lightweight, cost-effective, and easy on wildlife. It works too. Once zapped, cattle usually don't go near an electric fence again. Alternative methods of control include herding by a human, an ancient activity, and single-strand electric polywire, which is temporary and highly mobile. In all cases, the goal is the same: to gain control over the timing, intensity, and frequency of the animal impact on the land.

Planned grazing has other names: timed grazing, management-intensive grazing, rapid rotational grazing, short duration grazing, pulse grazing, cell grazing, or the "Savory system"—after the Rhodesian biologist who came up with the basic idea.

Observing the migratory behavior of wild grazers in Africa, Allan Savory noticed that nature, often in the form of predators, kept herbivores on the move, which gives plants time to recover from the pressure of grazing. He also noticed that because herbivores tended to travel in large herds, their hooves had a significant ground-disturbing impact, which he observed to be good for seed germination, among other things (think of what a patch of prairie would have looked like after a million-head herd of bison moved through). In other words, plants can tolerate heavy grazing, but not continuously. The key, of course, was that the animals moved on, and didn't return for the remainder of the year.

Savory also observed that too much rest was as bad for the land as too much grazing, meaning that plants can "choke" themselves with abundance in the absence of herbivory and fire, which remove old and dead material, prohibiting juvenile plants from getting established (not mowing your lawn all summer is a crude, but apt, analogy). In dry climates one of the chief ways grass gets recycled is through the stomachs of grazers, such as deer, antelope, bison, sheep, grasshoppers, or cattle. Animals, of course, return nutrients to the soil in the form of waste products. Fire is another way to recycle grass, though this can be risky business in a drought. If you've burned up all the grass, exposing the soil, and the rains don't arrive on time, you and the land could be in trouble.

The bottom line of Savory's thinking is this: animals should be managed in a manner consistent with nature's model of herbivory.

David and Kay James did precisely that—they adopted a planned grazing system for both their private and public land operations. And they have thrived ecologically and economically as a result. They saved the ranch too, and today the James Ranch is noteworthy not only for its lush grass and busy water, but for its bucolic landscape in a valley that is dominated by development.

David and Kay insist, however, that adopting a new grazing system was only part of the equation, even if it had positive benefits for their bank account. The hardest part was setting an appropriate goal for their business. This was something new to the Jameses. As David noted wryly: "We really didn't have a goal in the early days, other than not going broke."

To remedy this, the entire James clan sat down in the early 1990s and composed a goal statement. It reads:

> The integrity and distinction of the James Ranch is to be preserved for future generations by developing financially viable agricultural and related enterprises that sustain a profitable livelihood for the families directly involved while improving the land and encouraging the use of all resources, natural and human, to their highest and best potential.

Today David profitably runs cattle on 220,000 acres of public land across two states. He is the largest permittee on the San Juan National Forest, north and west of town. Using the diversity of the country to his advantage, David grazes his cattle in the low (dry) country only during the dormant (winter) season; then he moves them to the forests before finishing the cycle on the irrigated pastures of the home ranch.

That's enough to keep anybody incredibly busy, of course, but David complicates the job by managing the whole operation according to planned grazing principles. Maps and charts cover a wall in their house. But David doesn't see it as more work. "What's harder," he asked rhetorically, "spending all day on horseback looking for cattle scattered all over the county, like we used to, or knowing exactly where the herd is every day and moving them simply by opening a gate?"

It's all about attitude, David observed. "It isn't just about cattle." he said, "It's about the land. I feel like I've finally become the good steward that I kept telling everybody I was."

Recently, the family refined their vision for the land and community one hundred years into the future. It looks like this:

- "Lands that are covered with biologically diverse vegetation"
- "Lands that boast functioning water, mineral, and solar cycles"
- "Abundant and diverse wildlife"
- "A community benefiting from locally grown, healthy food"
- "A community aware of the importance of agriculture to the environment"
- "Open space for family and community"

And they have summarized the lessons they have learned over the past dozen years:

- "Imitating nature is healthy."
- "People like to know the source of their food."
- "Ranching with nature is socially responsible."
- "Ranching with nature gives the rancher sustainability."

But it wasn't all vision. It was practical economics too. For example, years ago David and Kay told their kids that to return home each had to bring a business with him or her. Today, son Danny started and manages a successful artisanal dairy operation on the home ranch, producing fancy cheeses for local markets; son Justin owns a profitable barbeque restaurant in Durango; daughter Julie and her husband John own a successful tree farm on the home place; and daughter Jennifer and her husband grow and sell organic vegetables next door and plan to open a guest lodge across the highway. Only one child, Cynthia, has flown the coop.

In an era when more and more farm and ranch kids are leaving home, not to return, what the James clan has accomplished is significant. Not only are the kids staying close, they are diversifying the ranch into sustainable businesses. Their attention is focused on the New West, represented by Durango's booming affluence and dependence on tourism.

Whether it is artisan cheese, organic produce, decorative trees for landscaping, or a lodge for paying guests, the next generation of Jameses has their eyes firmly on new opportunities.

This raised a question. The Jameses enjoy what David calls many "unfair advantages" on the ranch—abundant grass, plentiful water, a busy highway right outside their front door, and close proximity to Durango—all of which contribute to their success. By contrast, many ranch families do not enjoy such advantages, which made me wonder: beyond its fortunate circumstances, what can the James gang teach us?

I posed the question to David and Kay one evening.

"The key is community," said Kay. "Sure, we've been blessed by a strong family and a special place, but our focus has always been on the larger community. We're constantly asking ourselves, 'What can we do to help?'"

Answering their own question, David and Kay James had decided ten years earlier to get into the business of producing and selling grass-fed beef from their ranch—to make money, of course, but also as a way of contributing to the quality of their community's life.

Grass-fed, or "grass-finished" as they call it, is meat from animals that have eaten nothing but grass from birth to death. This is a radical idea because nearly all cattle in America end their days being fattened on corn (and assorted agricultural byproducts) in a feedlot before being slaughtered. Corn enables cattle to put on weight more quickly, thus increasing profits, while also adding more "marbling" to the meat, creating a taste that Americans have come to associate with quality beef. The trouble is cows are not designed by nature to eat corn, so they require a cornucopia of drugs to maintain their health. This is important because, as grass-fed advocate Jo Robinson puts it, "If it's in the feed, it's in the food." Meaning, it's in *you*.

There's another reason for going into the grass-fed business: it is more consistently profitable than regular beef. That's because ranchers can direct market their beef to local customers, thus commanding premium prices in health-conscious towns such as Durango. It also provides a direct link between the consumer and the producer, a link that puts a human face on eating and agriculture.

For David and Kay this link is crucial; it builds the bonds of community that hold everything together. "When local people are supporting local agriculture," said David, "you know you're doing something right."

Every landscape is unique, and every ranch is different, so drawing lessons is a tricky business, but one overarching lesson of the James Ranch seems clear: traditions can be strengthened by a willingness to try new ideas. Later, while thumbing through a stack of information David and Kay had given me, I found a quote that seemed to sum up not only their philosophy, but also that of the New Ranch movement in general. It came from a wall in an old church in Essex, England:

> *A vision without a task*
> *Is but a dream.*
> *A task without a vision*
> *Is drudgery.*
> *A vision and a task*
> *Is the hope of the world.*

The Allen Ranch *South of Hotchkiss, Colorado*

Stand on the back porch of Steve and Rachel Allen's home on the western edge of Fruitland Mesa, located 150 miles north of the James Ranch, in the center of Colorado's western slope, and you will be rewarded with a view of Stegnerian proportions: Grand Mesa and the Hotchkiss valley on the left, the rugged summits of the Ragged Mountains in the center, and on the right the purple lofts of the West Elk Mountains, a federally designated wilderness where Steve conducts his day job. Like the James Ranch, the Allens are permittees on the national forest, but what makes them unique is *how* they graze on public land: they herd.

I met Steve three years earlier at a livestock herding workshop I organized at Ghost Ranch, in northern New Mexico. I knew that his grazing association, called the West Elk Pool, had recently won a nationwide award from the Forest Service for its innovative management of cattle in the West Elks. The local Forest Service range conservationist,

Dave Bradford, had won a similar award for his role in the West Elk experiment. Intrigued, I invited them both down to speak about their success.

Steve began their presentation that day with a story. Driving to the workshop, he said, he and Dave found themselves stuck behind a slow-moving truck on a narrow, winding road. At first they waited calmly for a safe opportunity to pass, but none appeared. Then they grew impatient. Finally, they took a chance. Crossing double yellow lines, they hit the accelerator and prayed. They made it; luckily there had been nothing but open road ahead of them, he said.

The story was meant as a metaphor, describing Steve's experience as a rancher and Dave's experience with the Forest Service. The slow-moving obstacle, of course, was tradition.

In the mid-1990s, Steve and Dave convinced their respective peers to give herding a chance in the West Elks. They proposed that six ranchers on neighboring allotments, each of whom ran separate operations in the mountains, combine their individual cattle herds into one big herd and move them through the wilderness in a slow, one-way arc. By allowing cattle to behave like the roaming animals that they are (or used to be), Dave and Steve argued, the plants would be given enough time to grow before being bitten again, which in the case of the cattle of the West Elk Pool wouldn't be until the following summer.

There were other advantages, as they discovered. From the Forest Service's perspective, having one herd on the move in the West Elks rather than six relatively stationary herds was attractive for ecological and other reasons, including reduced conflicts with wildlife. For the ranchers, one big herd cut down on the costs of maintaining fences and watering troughs. It was also less labor-intensive, though it didn't seem that way initially. On the first go-around, Steve recounted, they had twenty people working the herd, which proved to be about twelve too many. Today, they move the herd with two to six people, and a bevy of hard-working border collies.

This was unusual because it is customary practice for ranchers to spread their cattle out over a landscape, especially in times of drought, not bunch them up. It's the "Columbus" school again—less management is the norm, not more. And herding means more management. Herding is also different because it is less dependent on *things*—fences, troughs, and other infrastructure—and more dependent on *people*. Not only is it an ancient human activity (think Persian nomads), but it dominated the early days of the Old West as well (think *Lonesome Dove*). Herding faded away, however, with the arrival of the barbed wire fence and, later, the federal allotment system for grazing, both of which splintered the wide open West into discrete units that lent themselves to a less intensive management style.

Dave and Steve had turned the clock back—or forward, depending how one looked at it. At the workshop they described the pattern of the herd's movement as looking like a large flowing mass, with a head, a body, and a tail, in almost continuous motion. Pool riders don't push the whole herd at once; instead, they guide the head—or the cattle that like to lead—into areas that are scheduled for grazing. The body follows, leaving only the stragglers—those animals who always seem to like to stay in a pasture, to be pushed along.

The single herd approach allows the permittees to concentrate their energies on all of their cattle at once, they reported, and also allows the Forest Service to more easily monitor conditions on the ground. Monitoring is often done in transects along the ground, a hundred feet at a time, say, usually starting at a permanently fixed point. Typically, the recorder takes measurements of plant diversity and spacing, soil conditions, and other indicators of range health.

In fact, the monitoring data showed such an improvement in the health of the land over time that the West Elk Pool asked for, and was granted, an increase in their permitted cattle numbers from the Forest Service. In other words, because the data supported their contention that herding was improving the health of land that had been beaten up historically by livestock, the ranchers of the West Elk Pool felt it was time to gain financially from their good work. This was significant because the trend in cattle numbers on public land was mostly the other direction—down—for a variety of reasons, not the least of which was pressure from environmentalists who saw cattle as simply destructive of land. But the West Elk Pool was different in this regard as well. They involved a local environmental group in the planning process and ultimately got its blessing for the herding experiment.

It was a strategy that paid off, literally.

Once again, as with David and Kay James, it started with a vision. After the workshop, Dave sent me the goal statement for the West Elk allotment, which reads in part:

> Our goal is to maintain a safe, secure rural community with economic, social, and biological diversity . . . that respects individual freedom and values education, and that encourages cooperation. . . . Our goal is to have a good water cycle by having close plant spacing, a covered soil surface, and arable soils; have a fast mineral cycle using soil nutrients effectively; have an energy flow that maximizes the amount of sunlight converted to plant growth and values the seclusion and natural aesthetics of the area.

Standing on the Allens' back porch, I asked Steve the question that had been on my mind since the workshop: what set him up for crossing those double yellow lines? A slight, quiet, and affable man, Steve didn't strike me at first as the ringleader type. Spend time with him, however, especially as he gently but firmly works his beloved border collies—he is a well-known trainer in the area—and you get the sense that a strong will is at work. Still, what leads someone to step out of the box like that?

Steve grew up in Denver, he said, where his father was an insurance salesman. He met Rachel at Western State College in Gunnison, where they discovered that they both liked to ski—a lot. Steve joined the ski patrol in Crested Butte, and eventually both of them became ski instructors. It was 1968. They were young and living the easy life. But restlessness gnawed at Steve. "The ski industry is designed to make ski bums, not professionals," he said with his easy smile. "It was fun, but we wanted more."

They were also restless about the changes happening in Crested Butte. Even in those early days, signs of gentrification were visible in town. Although not yet affected by the scale of change that transformed nearby Aspen into a playground for the rich and famous—a process sometimes called the "Aspenization" of the rural West—Steve and Rachel could see the handwriting on Crested Butte's wall. By the early

1970s, they decided to join the back-to-the-land movement, trading their skis for farm overalls.

"We weren't hippies, mind you," interjected Rachel, laughing. "We took farming seriously. I just want to get that on record."

They moved west, over the West Elks, to the small village of Paonia, where they planted what eventually became a large garden. They grew vegetables, raised chickens, produced hay, and learned from their farm neighbors.

"Because we admitted we didn't know very much," said Steve, "and because we were willing to learn, people were willing to teach."

This could be a motto for the New Ranch.

In 1977, restlessness struck again. They traded the garden for a run-down farm on the edge of Fruitland Mesa, where the hay was so bad the first few years they had to give it away. Eventually, they bought a few cattle and decided to try their hand at ranching. In 1988, Steve purchased a Forest Service permit in the nearby West Elks, mostly as a forage reserve for his animals in times of drought. His interest was not purely economic, however. Steve had always been attracted to mountains, and soon he had a chance to work in them daily.

Eager to learn more, Steve took a Holistic Resource Management course the same year that Dave James did and that led him to give herding a try. With the arrival of Dave Bradford in the Forest Service office in Paonia a short while later, the opportunity to cross the yellow lines suddenly presented itself.

As part of the process of pulling the West Elk experiment together, Steve also became a student of a new method of low-stress livestock handling sometimes called the "Bud Williams school," after its Canadian founder. Its principles fly in the face of traditional methods of cattle handling, which are full of whooping, prodding, pushing, and cursing. Putting stress on cattle is as customary to ranching as a lasso and spurs.

But that was Steve's point: customary, yes; natural, no. And that's where herding comes in: pressure from predators in the wild made grazers naturally bunch in herds. Unfortunately, on many ranches today, the herd instinct has been prodded out of most cattle.

The whole idea of low-stress handling is to use a "law of nature" to positive effect.

"Nature," Steve said simply, "has the right ideas, but we keep messing them up."

It is this return to what's seen as closer to nature's original model, such as grass-fed livestock and low-stress herding, that defines the progressive ranching movement underway today.

But the New Ranch is also about relationships, especially among teachers and students, mentors and apprentices. Learning should be a lifelong affair, which means students of all stripes need teachers—people like Steve Allen. Grass may lie patiently for rain, but people need inspiration.

Twin Creek Ranch *South of Lander, Wyoming*

A few hours north of Fruitland Mesa I entered the dry heart of that summer's drought, which was centered on southern Wyoming. After crossing Sweetwater River on the old Mormon Trail, I knew the precise moment when I had reached Tony and Andrea Malmberg's ranch. Rounding a big bend in the road, I was suddenly confronted with the sight of green grass, tall willows, sedges, rushes, and flowing water.

I had arrived at Twin Creek. Following the lush riparian strip toward the ranch headquarters, I recalled an anecdote Tony had included in an article titled "Ranching for Biodiversity," which he had recently written for The Quivira Coalition's newsletter. It detailed an experience of his youth when he and a brother-in-law decided to blow up a beaver dam on the creek:

> Jim and I crawled through the meadow grass under his pickup giggling. Jim pulled the wires in behind him, leading to the charge of dynamite.
>
> "This will show that little bastard," I said. Jim touched the two wires to the battery. WOOMPH! The concussion preceded the explosion. Sticks and mud came raining down on the pickup. As soon as it stopped hailing willows and mud, we scrambled out from under our shield.
>
> "Yeah!" I hollered as we ran down the creek bank. "I think we got it all."

Water gushed through the gutted beaver dam and we could see the level dropping quickly. The next morning I rode my wrangle horse across the restored crossing. The water behind the beaver dam had gotten so deep I couldn't bring the horses across. But that was taken care of now. I galloped down the creek. The water ran muddy and I couldn't help but notice creek banks caving into the stream.

I wondered.

It was another story about tradition, this time about conventional attitudes toward wildlife. But it was also an allegory. Years later, when Tony was a young man, his family ranch "caved in" too—forced into bankruptcy by high interest payments on loans and tumbling cattle prices, costing Tony's family the entire 33,000-acre property. Suddenly homeless, Tony began to wonder what had happened. Two years later, he leased the ranch back from the new owner, before eventually buying it. But he knew things had to be different this time if he wanted to stay.

Like David James and Steve Allen, Tony attended a course on Holistic Resource Management, and he began to realize that biodiversity was a plus on his ranch, not a minus. "I shifted my thought process to live with the beaver and their dams," he wrote in his article. "With this commitment, I viewed the creek as a fence rather than something I could cross. This attitude gave me an extra pasture, a higher water table, less erosion, and more grass in the riparian area. The positive results energized me, and I began to curiously watch in a new way."

What he noticed as a result of his new land management was an increase in biodiversity. Moose, previously a rare sight on the property, began to appear in larger numbers. He even began to appreciate the coyotes and prairie dogs on the ranch and the role they played in the health of his land. Later, a University of Wyoming study found a fifty percent increase in bird populations over the span of a few years.

All of which led him to formulate two guiding principles:

First, I avoid actively killing anything, and notice what is there. Whether a weed or an animal, it would not be here if its habitat were not. I plan the timing, intensity, and frequency of tools (grazing, rest, fire, animal impact, technology and living organisms) to move community dynamics to a level of higher diversity and complexity.

Second, I ask myself what is missing. Problems are not due to the presence of a species but rather the absence of a species. The absence of moose meant willows were missing, which meant beaver were missing and the chain continues.

If I honor my rule of not suppressing life, I will see beyond symptoms to address problems. If I continue asking "What is missing?" I will continue to see beyond simple systems and realize the whole. When I increase biodiversity I improve land health, I improve community relations, and I improve our ranch profitability.

To accomplish his goals, Tony employs livestock grazing as a land management tool. To encourage the growth of willows along the stream and ponds, for example, he grazes them in early spring to assist seedling establishment. By concentrating cattle for short periods of time in an area, Tony breaks up topsoils and makes the land more receptive to natural reseeding and able to hold more water.

What brought me to Twin Creek, however, wasn't just the tall grass, the flowing water, or even the progressive ranch management practiced by Tony and Andrea, though these were important. What I wanted to see was the very nice bed and breakfast they operated.

As I pulled up to the spiffy, new three-story lodge, I was greeted with a sunny wave by Andrea. A child of the Wyoming ranching establishment—her father traded cattle for a living—Andrea heard Tony speak passionately some years earlier about the benefits of planned grazing at a livestock meeting (where his talk was coolly received) and wrote him an equally passionate letter challenging his beliefs. They corresponded back and forth until she accepted his dare to come to the ranch and see the proof herself.

Tony joined us inside the airy lodge. Bearded, deep-chested, and sporting a leather vest, Tony looks the part of the cowboy. He is also cheery and garrulous, in print and in person.

Over a glass of wine later that evening, I learned that the lodge is the happy ending to a story that had its roots in anger. "When my family lost the ranch," recalled Tony, "I blamed everyone but ourselves. I blamed consumers, environmentalists, liberals. But most of all, I blamed our new neighbors."

In 1982, as the family was slipping into bankruptcy, a man from California bought a neighboring ranch for twice what a cow would generate per acre. Although this fact didn't directly affect his family's pending insolvency, it angered Tony because it suggested the end of an era. Ranch land had more value to society, he saw suddenly, as an amenity than as a working landscape. Recreation trumped ranching. The Old West was giving way to the New. And Tony didn't like it.

But then Tony had a revelation: markets don't lie. Upon returning to the ranch, he decided that in addition to the cattle operation he would start a ranch-recreation business and market overnight stays directly to people who wanted the cowboy experience. He quickly learned, however, that paying guests wouldn't tolerate dirt or mice as much as he did, so he and Andrea took the plunge and built a pretty lodge with a capacity for fourteen guests at a time.

But they didn't stop there. Making it economically meant exploring as many diverse business enterprises as possible. Andrea convinced Tony that the next step was to "go local" and find ways to tap nearby markets, including their new neighbors, for their beef and other services. They hosted a class on weed control for local ranchette owners and focused on the positive role of goats, which will eat every noxious weed on the state list. It was a big hit.

That was followed by a seminar on rangeland health, which proved popular with their ranching friends. Then came a foray into the grass-fed beef business, which has been successful too.

Next in their efforts at economic diversification was Andrea's decision to start teaching yoga. A recent winter solstice party packed the lodge with what Tony called the "strangest assortment of people I'd ever seen together."

"The hodgepodge appeared to be a demographic accident," he continued, "yet they all ended up in central Wyoming because they wanted the same things we want: a beautiful landscape, healthy ecology, wholesome food and a sense of community." In this, Tony drew a parallel with the benefit of increased biodiversity on the ranch.

"In the old days, I didn't have to deal with people different from me," he said. "But this is better."

Tony went on to explain to me how his indicators of success have changed over the years. In 1982, his primary measure of success was a traditional one: increased weaning weights of his calves. By 1995, Tony's measure had shifted to the stocking rate of cattle (the number of cattle per acre on the ranch: more cattle, managed sustainably, equals greater profitability), which, thanks to planned grazing, was up 75 percent from years prior. By 1998, his indicator had shifted to monitoring and what it said about trend. In his case, the trend was up, meaning his land showed increased plant health and decreased soil erosion. By 2000, Tony used the diversity of songbirds on the property as his baseline (over sixty species at last count). By 2002, however, the main measure of success had changed to an economic one: how many activities generated income for the ranch in a year. At the time of my visit, they were up to three.

Tony attributes this success to their ability to speak different languages to different audiences, including recreationalists.

"I realized that if I'm going to survive in the twenty-first century, I need to be trilingual," Tony explained. "Ranchers tell stories. The BLM [Bureau of Land Management] wants to talk data. And then we've got the environmentalists. Lander has a lot of them. To connect with them you need to use poetry."

In other words, success in ranching today is as much about communication and marketing as it is about on-the-ground results. As Tony and Andrea's story suggests, it is not enough simply to *do* a better job environmentally, even if it brings profitability. One must also *sell* one's good work, and do so aggressively in a social climate of rapid change and the general population's increasing detachment from our agricultural roots.

From all the indicators that I saw, Tony and Andrea are on the right track. The lodge was clean, comfortable, and airy; the food was wonderful; and the visitors were happy. But this is no dude ranch. Tony makes his guests work. According to his planned grazing schedule, his cattle need to be moved almost every day, so he has paying guests do it. They love it, of course, and because his cowboy does the supervisory work, Tony is free to explore other business ideas. And the ideas keep coming.

Red Canyon Ranch *West of Lander, Wyoming*

When I met Bob Budd at The Nature Conservancy office in Lander, a short drive north from the Malmberg's Twin Creek Ranch, he was pacing the floor, waiting for my arrival.

"The ranch is on fire," he said quickly. "Let's go."

And go we did. Despite being a foot taller than Bob, I had to hustle to keep up with him as we headed outside. A Wyoming native son, a member of a well-known ranching family, and former executive director of the state's cattlemen association, Bob managed the Red Canyon Ranch for The Nature Conservancy's Wyoming office when I met him. He also served as their director of science. Bob had earned a master's degree in ecology from the University of Wyoming and was in line to become president of the Society for Range Management, a highly respected national association of range professionals.

Without a doubt, he was a man on the move.

I jumped into my truck and followed Bob rapidly to the head-quarters of the thirty-five-thousand-acre Red Canyon Ranch, which borders Lander on the south and west. The Nature Conservancy, Bob said, had purchased the property for three reasons: to protect open space and the biological resources held there; to demonstrate that livestock production and conservation are compatible; and to work at landscape-level management and restoration goals.

The first two goals have more or less been achieved, he said as I climbed into his truck in the ranch's parking lot. It is the third goal that motivates him now. What Bob wants is fire back on the land, brush and trees thinned, erosion repaired, noxious weeds eradicated, perennial streams to flow fuller, riparian vegetation to grow stronger, and wildlife populations to bloom.

And judging by the speed at which we traveled, he wanted them all at once.

Bob was thrilled about the lightning-sparked fire that was burning a chunk of forest and rangeland right where he had been encouraging the Forest Service to light a prescribed burn for years. That's because fire is a keystone ecological process, meaning a process that is fundamental to the health of the ecosystem over time. Research shows that "cool" fires happened frequently in western forests, perhaps as often as every ten years in some stands. But for much of the twentieth century, humans suppressed all fires in our national forests, mostly to protect the monetary value of the timber, and as a consequence the forests have become overgrown and dangerously prone to "very hot," destructive fires, as we saw in 2002. To reverse this condition and restore forest health, ecologists and others have encouraged the Forest Service to light controlled burns. To many, however, the pace of bureaucracy has been frustratingly slow.

"I love lightning," Bob said with a twinkle in his light blue eyes, "because there's no paperwork."

As we sped into the mountains in search of a suitable vantage point to observe the progress of the fire, talking energetically about ecological theories that I had only recently begun to study, I recalled Bob's essay in a book titled *Ranching West of the 100th Meridian.* He wrote: "I am an advocate for wild creatures, rare plants, arrays of native vegetation, clean water, fish, stewardship of natural resources, and learning. I believe these things are compatible with ranching, sometimes lost without ranching. Some people call me a cowboy. A lot of good cowboys call me an environmentalist."

Bob has strong words for both, especially about their respective defense of myth. He likes to remind environmentalists in particular that nature is not as pristine as many assume. For thousands of years, he observed, Wyoming has been grazed, burned, rested, desiccated, and flooded. In saying so, he consciously tilts at an ecological holy grail called the "balance of nature." This is the long-standing theory that says nature tries hard to hold things in balance; in other words, when a system gets "out of balance," nature works to right the ship, so to speak. Predator/prey populations are a good example. According to this theory, too many coyotes and not enough jackrabbits, say, mean nature will bring the coyote population back into balance over time (by starvation).

Today, most professional ecologists reject this theory in favor of one called the "flux of nature," which views nature as dynamic, chaotic, and rife with bouts of disturbance, such as forest fires and floods.

Unfortunately, the "balance of nature" theory persists among many nonprofessionals, especially nature enthusiasts.

"In landscapes where the single ecological truth is chaos and dynamic change," Bob wrote, "we seem obsessed with stability. Instead of relishing dynamic irregularities in nature, we absorb confusion and chaos into our own lives, then demand that natural systems be stable."

He likes to explain to both environmentalists and ranchers that grazing, like fire, is a keystone process. "Like fire, erosion, and drought, grazing is a natural process that can be stark and ugly," he wrote. "And, like fire, erosion, and drought, grazing is essential to the maintenance of many natural systems in the West. . . . Adults tend to overlook other grazing creatures, we forget the impact of grasshoppers, rodents, birds, and other organisms that have long shaped the West."

Just as prescribed fire, once controversial, is now widely accepted, Bob observed, it is simply a matter of time before the same change of thinking happens to grazing.

Mulling this thought over as we sped through the forest, still searching for a spot to view the fire, I asked him if he thought environmentalists would ever embrace ranching.

"I think they'll have to," he replied, "if they want to protect open space."

Bob explained that in Wyoming, like much of the West today, unbridled development on private land has resulted in habitat fragmentation and destruction. When land is subdivided, the new roads and homes often interrupt wildlife migration corridors, decrease habitat for rare plants and animals, and make ecosystem management difficult. The open space ranches provide is the last barrier to development in many places. "The economic viability of ranching is essential," he said, "in maintaining Wyoming's open space, native species, and healthy ecosystems."

"Even on public lands?" I asked.

"Absolutely," he replied. "That's because it's all about proper stewardship. I don't care where you are."

Bob pointed to a thick stand of trees outside the truck window.

"Our common goal must be to provide the full range of values and habitat types that a variety of species need, including us," he said. "And ranchers can help." Ranchers can become restorationists, he continued, because they are uniquely positioned to deliver ecological services—food, fuel, fiber, and other ecological benefits that society requires—as landowners, as livestock specialists, and as hardworking sons- (and daughters) of-guns.

This will become increasingly important, I'm convinced, as the twenty-first century wears on and we come to realize just how much restoration work is required—not to restore the "balance of nature" but to get nature back into a position where it can operate according to natural principles, including disturbance. Cows can have a role here too. As domesticated animals they can be used effectively to recreate certain kinds of animal impact on the land—a point Allan Savory made years ago.

Suddenly, we stopped. The fire we sought had proved elusive, and it was time to head back to headquarters. It seemed symbolic. Although landscape-scale opportunities for ranchers may be plentiful, as Bob suggested, many are elusive, especially on public land, where every action seems to engender an opposite reaction by someone. Even the smallest restoration project, whether it involves livestock or not, can quickly become mired in red tape and conflict. Bob remained optimistic, however. He admitted that he had to be.

Returning to the ranch, Bob kept moving. He needed to take his son to baseball practice. I followed him into the house for introductions to the family. We talked for a while longer, shook hands, and before I knew it, he was gone.

Rather than drive off immediately too, I walked down to a bridge that spanned a burbling creek. Enjoying a momentary respite from the dust, driving, and cascade of ideas, goals, and practices that had dominated conversation for the entire trip, I leaned on the wooden railing and listened to the wind.

A variety of threads, I saw, tied Bob Budd to the Jameses, Allens, and Malmbergs. One was the desire to make amends with nature. To paraphrase President John Kennedy, each asked "not what the land could do for them, but what they could do for the land." Whether it was restoring land to health, bridging urban-rural divides, teaching, feeding, or peace making, every person I encountered was engaged in an act of redemption, mostly by trying to heal damaged relationships, particularly our bond with the land. This is good news for grass, especially in these dry times. It is probably good news for us all as well.

Which led to a second thread: grass may seem immortal, but in reality it needs water, nutrients, animals, and fire to stay vigorous. The health of the whole depends on the health of its essential parts. This is important, as Bob Budd explained, because disruption is inevitable in nature; sooner or later, a calamity of some sort will strike and those plant and animal populations that are not functioning properly at basic levels will be in jeopardy. Communities of people are no different. Whether it is a ranch, village, small town, or city, every community needs to be diverse, resilient, opportunistic, and self-reliant if it is to survive unexpected challenges.

For example, by setting water to work with a purpose—to earn a living within nature's model—the James family has buffered themselves well against uncertainty, and in the process protected 400 acres of prime land along the Animas River from subdivision. The potential financial gain from busting their land into small lots for houses is astronomical, but they won't do it, because it doesn't fit their goal for their family, their land, or their community.

Or take Steve Allen. Not only did he come to realize that herding is an ancient form of sustainability, he took the unusual step of crossing double yellow lines to achieve his goal. How many of us city folk are willing to take a risk like that? Do we even know where the yellow lines are? Are *we* resilient in our own lives? Or are we in spiritual (as well as practical) danger of supposing, as Aldo Leopold warned, that "breakfast comes from the grocery, and heat comes from a furnace."

A third thread was Tony Malmberg's question about the sanctity of life—when might we stop killing things we don't understand, as he did, and start inquiring instead about what might be missing from our lives? And once the outlines of answers become perceptible, what language do we speak so the lessons we've learned can be clearly understood?

Must we be trilingual, or at some point will one vocabulary suffice—the language, say, of grass? Or food? And if we can figure all that out, how do we make it *pay*—as in paychecks—without which little can be accomplished?

I leaned wearily on the railing above the burbling creek.

A fourth thread involves the Big Picture. How do we work at scale, as Bob Budd advocates, and not just on ranches and farms, but all over the West, the nation, the globe? How do we take a landscape perspective in a world balkanized into countless, and often feuding, private, state, tribal, and federal fiefdoms? How do we overcome the paperwork, the lawsuits, the power struggles, and the politicking necessary to get the big work done in a century that will likely be roiled by climate change, energy instability, water shortages, and a host of other potential challenges?

I wasn't sure I had a clue. No wonder humankind invented alcohol.

But I did have a clue, of course. I found it at nearly every stop along this trip. Stegner was right, ranching *is* renewable; in fact, it feels very much like it's being reborn, one ranch at a time. And in an era dominated by disposability and depletion, this is important news. Grass and grazers, after all, are the original solar power. Moreover, humans have been living and working with livestock for a very long time and through a great deal of historical change. The human desire to be near animals, and be outdoors, hasn't altered much over the centuries, though it has recently shrunk, hopefully temporarily, as a result of industrialization. We need ranching, I came away thinking, because it can be regenerative, not only for the food and good stewardship it can provide, but also for the lessons it can teach us about resilience and sustainability. All flesh is grass, as the Bible reminds us, though this has often been forgotten.

Perhaps it was time to consider it again.

Chapter Two

REPLENISHING SOIL AND PEOPLE

Carrizo Valley Ranch *North of Capitan, New Mexico*

If land management is more art than science, as many say it is, then Sid Goodloe will enter the history books as one of the West's great artists and his Carrizo Valley Ranch, located at the upper end of a lovely valley in mountainous central New Mexico, a masterpiece.

This idea came to me one day while standing on a hill on Sid's ranch, surveying the lush savanna expanses on the eastern half of his property and the parklike ponderosa pine forest to the west. I suddenly realized that I was looking at a work of art, fifty years in the making. It had all the hallmarks of a great piece of sculpture: beauty, proportion, vitality, skill, design, and effect. That the chosen medium was land, not marble, made no difference, the result was the same: awe and inspiration.

Only don't call Sid Goodloe an artist. Blue-eyed, gray-bearded, cowboy-hatted, and dressed in a plaid western shirt, blue jeans, and faded cowboy boots, Sid is a rancher first and last, with work to do, as we did that day. In fact, my momentary reflection happened only because Sid had walked off, gas can in hand, to inspect a pile of juniper trees that he had recently cut and pushed into a heap.

"It's a good day to burn," he had announced at sunrise. And true to his word, a few minutes after we left the truck the pile erupted into flames.

As I watched the fire grow, thinking now about performance art, Sid returned to the truck, limping slightly—at seventy his only concession to age being hearing aids and a bum knee. He had no reason to be impressed by the conflagration he had just created; he had lit similar fires a thousand times by now. He tossed the gas can into the back and climbed into the front seat without a backward glance.

"Let's go," he said in his Texas drawl. And we went.

As we bounced over the rutted road, I pondered my revelation. Like all great artists, Sid is visionary, intuitive, opinionated, curious, driven, exacting, and boundlessly energetic. But he is also pragmatic, flexible, and results-oriented. He has to be, or he wouldn't be in ranching for very long.

Perhaps that's what makes him and his work so inspirational; in blending art with science, vision with profit, and opinion with outreach he has become—by near unanimous acclaim—one of the preeminent land managers of our time. And his success rests on a simple foundation: he learned to bend his economic needs to the needs of the land, becoming, in the process, a living example of Wendell Berry's famous maxim that sustainable agriculture is that "which depletes neither soil nor people."

He did so as a modern-day pioneer, trailblazing a creative career with a gas can, matches, and cattle, instead of brushes and oils. And in the Carrizo Valley ranch he found the perfect canvas—a *depleted* canvas, as I found out.

In the early 1950s, after an education at Texas A&M and a stint in the army, Sid left his west Texas homeland to cowboy on what were then big ranches outside Roswell, in eastern New Mexico. As he worked he also searched for, in his words, "a worn-out piece of ground"—affordable land—that would be amenable to multiple use, had potential for profit, and was a good place to raise a family.

In 1956, Sid found what he was looking for in the mountains near Capitan. It was Billy the Kid and Smokey Bear country—two characters that would play prominent roles in his life before long. What he found were 3,500 acres of dusty, scrubby land surrounded on three sides by the Lincoln National Forest, choked at its higher, western end with spindly trees, and split down the middle by a dry, rocky streambed.

Thinking that this was what New Mexico was "supposed to look like," Sid purchased the property and began trying to make a living on a landscape that, he quickly learned, had been battered and bruised. Continuous year-round grazing since the pioneer days had stripped the land of health. Additionally, the nationwide Forest Service policy of "all fires out by 10 a.m." had caused thick stands of trees to invade open spaces on areas of the ranch previously kept clear by fire from the neighboring forest. As a consequence, Carrizo Creek had dried up, wildlife was scarce, and the carrying capacity was only sixty cows.

It was an Old West legacy typical of the entire region.

"I paid $19 an acre for the land, and for a while I thought I had been taken," he told me. "I had no idea that the massive number of livestock that came west after the Civil War and the good intentions of Smokey Bear were the cause of denuded watersheds and water shortages."

Although initially impressed by the toughness and self-reliance of his forebears and neighbors, eventually Sid realized that those characteristics rarely translated into proper land stewardship. Looking around, Sid understood that a fragile ecosystem had succumbed to the ignorance and economic hard times that came with the area's pioneer heritage.

"It wasn't their fault, they were acting on the best knowledge they had at the time," he said, "but it sure made a mess of the land."

Two clues turned Sid's thinking around. The first was his discovery in Carrizo Canyon of a prehistoric petroglyph of a fish, suggesting that the creek here once sustained aquatic life. The second occurred after he attempted, and failed, to locate the southwest corner marker of his property, which was buried in a dense thicket of trees. In frustration, he dug up a survey report from the 1880s and was astonished to read that the surveyors couldn't find a tree to mark anywhere near the corner position.

"Obviously something had gone very wrong," recalled Sid. "So I began to count the rings on the ponderosa trees that grew on the ranch.

Ninety percent of those plants had come up after the Civil War. It was then I began to visualize open woodland, savanna, and grassland."

It was then that the ranch began to be the raw material for a masterwork.

"Why not try to return to those days of stirrup high grass and streams with beaver dams and cutthroat trout?" he wondered.

The first step would be getting the cattle part right. When he purchased the property, Sid implemented a classic rest/rotation system that he learned in college, dividing the ranch into summer and winter pastures, which was only a slight variation on the "Columbus school" of grazing management. It was the best he thought he could do for his sixty head of cattle. Also in accordance with his training, the goal of his brush control program was strictly to increase more grass for his cattle.

But if the plan was to restore the ranch to ecological health, it wasn't happening.

Conditions changed dramatically in the late 1960s, however, when Sid switched to a short-duration grazing system that he picked up from wildlife biologist Allan Savory himself, whom Sid met while touring what was then Rhodesia. In the process, Sid likely became the first rancher in the United States to implement such a system. He was certainly one of the first people to write on the topic when the *Journal of Range Management* published his article "Short Duration Grazing in Rhodesia" in 1969.

What was different, though, wasn't just the technical practice of rotational grazing—today the Carrizo Valley Ranch has twelve paddocks, seven for summer use, three for winter, and the other two for spring and fall. What was important was the way Sid began to look now at the land as a whole system, not just in terms of what was good for his cattle.

"My goal is an integration of all components—economic, human, and environmental," he wrote recently in a paper he gave me to read, "into a synergistic, comprehensive plan that allows management for long term sustainability rather than short term production." For example, Sid defers most of his riparian area from livestock grazing between May and November and only "flash grazes" it during the dormant season. The short-duration grazing plan gave his cool-season grasses (plants that grow primarily in spring or fall) a break, which had an economic benefit as well.

"Managing for cool-season plants has probably made me more money than any other practice," he told me. "Spring forages are simply a lot more valuable than summer forages and have more to do with a good calf crop than almost anything else."

Next, Sid switched to a breed of cattle he calls Alpine Black—a cross between Angus and Swiss Brown—that suited the mountainous environment better. Once again, his goal was to make his operation fit the land, not vice versa.

But his masterwork really began to take shape when he tackled the difficult and tenacious problem of too many trees.

"Everyone told me I was crazy," he recalled. "They assumed that the invading trees and brush was just the land returning to its natural state."

At first, he used mechanical thinning and chemicals to control what he perceived to be an overabundance of woody material on the property, but the key moment came when he realized that he needed to get fire back on the land.

"Tree ring studies in New Mexico show that most forests burned every seven to ten years," Sid said. "Whether it was Indians or lightning, the land was kept in an open savanna state. Fire is as important to the land as rain and sunlight. It is the natural predator of the forest." Sid realized that when the government stopped letting the forests burn, the trees grew with a vengeance and the whole system unraveled.

To achieve his vision for the land, Sid began an aggressive program of thinning and burning, earning the sobriquet "Pyro Goodloe" in the process. To help finance the restoration work, he harvested some of the wood for sale as firewood, fence posts, and vigas (beams that hold up the famously flat roofs of New Mexican homes) to the residents of the nearby and rapidly growing tourist centers of Ruidoso and Lincoln— home, briefly, to the outlaw Billy the Kid.

Today, cool-season grasses have returned in abundance, the brush and tree invasion has been turned back, water runs year-round in the creek (or at least in nondrought times), carrying capacity for livestock has increased by 30 percent, and wildlife is thriving. Visitors are often greeted by a pronghorn herd at the ranch's entrance. Some of this was the result of tree work, some of it cattle work, and some of it just plain elbow grease on the part of Sid. Of course, rain helps too.

But it all worked together because it was integrated under one vision.

"My ranch was the first holistic management venture in the country," he said matter-of-factly. "It took a lot of years and I made many mistakes. Now I have figured it out and I try to share that. I'm an environmentally sensitive rancher. It makes me money and I like it!"

Sid has done more than replenish a worn-out stretch of land, however. Through his tireless advocacy and outreach—he has led countless tours of his ranch over the years and spoken at countless conferences and meetings—Sid has shared his vision and his sense of hope with thousands of people, including many city-based conservationists. In the process, he has probably done more to repair the bridge between urban and rural residents than any other rancher in New Mexico, and possibly in the region. This is important because this bridge, strong up through mid-twentieth century, has weakened considerably recently. As sources of food, fuel, and fiber become increasingly abstract to urban residents, the bond between city and country has dissolved into an abstraction as well, leading to various forms of estrangement. One example is the grazing "debate" of the mid-1990s in which urbanites, flush with cheap food and fuel from distant lands, could literally afford to sever relationships with nearby working landscapes.

In an attempt to reach city folk with his message about ranching, Sid was a one-man band for a while, trying his best to keep the urban-rural bond from breaking entirely. It was for this reason, plus the healthy condition of his ranch, that when The Quivira Coalition organized its first on-the-ground workshop, way back in 1998, it was a no-brainer to ask Sid to host it at his place. We called it an Outdoor Classroom on Rangeland Health and our goal was to attract ranchers, conservationists, and other members of the public so they could shake hands and learn from one another. Sid proved to be the perfect host, too, for the twenty-five folks that showed up.

But I wanted to do more than build bridges with this workshop. I wanted the group to study the common ground below our feet—to talk the language of soil, grass, and water. That's why I asked Kirk Gadzia to teach the Classroom (the first of many he would do). I had met Kirk only a few months earlier and saw that he was a natural teacher. He also knew his stuff. As one of the coauthors of the National Research Council publication titled *Rangeland Health*, Kirk knew all about healthy land—water cycle, nutrient cycle, energy flow—and I suspected he could talk about it to a diverse audience in a non-wonkish way. Reading *Rangeland Health*, I also realized that the language of land health could be used to describe communities of people as well as communities of plant life.

As an illustration, here's a short primer on rangeland health from materials Kirk used in our Outdoor Classrooms—and as you read it think about how its words could describe *our* health as well:

- *An effective water cycle*: a permeable soil surface; evaporation losses minimized; the effects of floods and droughts are less severe; underground water levels are stable or easily replenished; soil organic content is high; plant growth rates are fast
- *An ineffective water cycle*: soil surface is sealed and crusted; evaporation is high; the effects of droughts and floods are severe; water levels are not easily replenished; soil organic content and plant growth rates are low
- *An effective mineral cycle*: deep roots; rich organics; high plant species diversity; rapid cycling of nutrients via plant roots.
- *An ineffective mineral cycle*: high degree of bare ground; high erosion; shallow roots; low biological activity
- *High successional communities*: Composed of populations of many different species of plants, animals, birds, insects, and microorganisms; not prone to wide fluctuations in populations or normal weather extremes
- *Low successional communities*: Composed of populations of only a few species, relative to the potential of the site; usually unstable and vulnerable to fluctuations and extremes in population shifts and weather

To explain these principles, the first thing Kirk did, after introductions, was lead us across one of Sid's big savannas to the boundary with the national forest. Guided by Sid, we crossed the fence and continued our stroll into a thick patch of juniper trees whereupon Sid signaled us to stop. Using a small shovel he dug a small trench in a large area of bare soil between two trees. Then he drenched the excavation with water from a backpack sprayer. We saw a million little juniper roots jutting from the sides of the trench. "That's why there's no grass here," Sid explained. "There's too many roots from too many trees!"

"As a society, we're quick to lay blame," added Kirk. "In this case, it would be easy to blame cattle for the lack of grass. But in reality, it's much more complicated than that. As it is with most things."

Deftly handling complicated relationships among people, land, and animals is the main reason why he may be responsible for the improvement of more acres of rangeland in the Southwest in the past two decades than any other single individual. For more than twenty-five years, Kirk has taught, consulted, or otherwise worked with ranchers and other landowners whose global combined holdings run into the multiple millions of acres. And if the feedback from his clients around the Southwest is any indication, most of those acres have become healthier as a result of his tutelage. That's not bad for a man who has never owned a cow or managed a ranch.

"Other than pounding a few posts, it's been mostly talk," said Kirk with his easy smile and twinkling blue eyes. "In fact, I like to think of myself as a 'professional visitor.'"

More often dressed in sandals, shorts, and a Hawaiian shirt than jeans or cowboy boots, and quick to share a bottle of home-brewed beer or salsa from his garden, Kirk's affability has been a key to his success in working with ranchers from Canada to Mexico to South Africa to Hawaii, the Midwest, Florida, Virginia, and Australia.

The thread connecting all his clients is not cattle, however, but land health.

"It has been a tremendous opportunity to see so many ecosystems," he said. "And what I've learned is no matter where you are on the planet, ecosystem functions are the same. Another thing I have learned is that people are a lot like ecosystems, they have boundaries and thresholds too," he continued. "It's something many people didn't understand about the Savory model—that it wasn't about cattle, it was about eco-system principles and about how to manage people with the similar kinds of principles."

In fact, in his experience the biggest obstacle to improving a ranch's economics or getting the range into better ecological shape is often con-tentious family dynamics, which is why Kirk does a lot of facilitation with family members before he gets to questions about plants and ani-mals. A measurement of Kirk's skills in the "people department" is that he doesn't work with contracts. And he's never had a check bounce.

Unlike many ranch consultants, Kirk does not have a background in agriculture, though he discovered a love of the natural world at an early age. Born in Tampa, Florida, where his father served in the air force, Kirk was drawn early to the outdoors, especially hunting. By the time he entered high school—his family then lived in the Mojave Desert—he spent most of his free time hiking in the Sierra Nevadas.

At Brigham Young University, he majored in wildlife biology, but minored in range. Curious about the connections between the two, he went on to earn a master's degree from New Mexico State University in range science. In between he worked for the Bureau of Land Manage-ment doing vegetative and wildlife inventories, including field botany (he's an expert at plant identification), which also satisfied his desire to be in remote places.

It was only during his master's degree work that he began to think about cattle. "I thought I didn't have a prejudice against cows when I began my research," he told me, "but when I began to study black grama grass I found myself automatically assuming that the study plots con-tained in exclosures would be healthier." To his surprise, they weren't.

"I saw better establishment of black grama grasses in grazed areas," he continued. "But the real revelation came when I learned that the exclosures were mesquite-free not because they were cattle-free but because they had been sprayed by herbicide. Everything changed after that."

Upon graduation, he moved to Albuquerque and took a job as a range conservationist with the Bureau of Indian Affairs' Southern Pueblos Agency. The work was stimulating and educational, and he enjoyed working with the tribes, though not the endless meetings it entailed.

In 1980 his life changed when he took a two-week course taught by Allan Savory and Stan Parsons. Both men had recently relocated to Albuquerque from Rhodesia, bringing with them a new vision of people and land—a vision that was considered heretical by many in the range and ranch communities because it ran counter to the traditional school of ranch management that believed that dispersal of cattle, not bunching them up, was the proper way to graze.

"It was an era of great excitement in range management," Kirk recalled. "I was thrilled to be a part of it."

With the assistance of Savory and others, Kirk convinced Sandia Pueblo, a Native American tribe located just north of Albuquerque, to install a "Savory" cell system on twenty thousand acres of tribal land. This system resembled a pie, with a water source at the center of the pie and "wedges" constructed of electric fencing running outward. The idea was to increase the number of paddocks significantly so that tribal ranchers could control the timing, intensity, and frequency of their cattle on the land. Kirk wrote the plan, directed the fence building, and did the monitoring.

"A lot of tribal members came out when we started," he recalled. "And it worked well for a while. But eventually it failed, and it failed because it was our project, not theirs. They didn't have ownership. They chose not to fix the fences after we left, for instance."

The experience taught Kirk a valuable lesson: that the people part of land management is as important, if not more important, than the stewardship toolbox.

Eventually, Kirk left the Bureau of Indian Affairs to work for Savory's nonprofit organization as its education director. During the next seven years he taught a holistic management course once a month to twenty or more ranchers and others around the region. One of his early students was Jim Winder, one of the cofounders of The Quivira Coalition. In 1994, after honing his teaching skills, and touching a lot of lives, Kirk left the nonprofit world to start his own business. One of the lessons he took with him was to never work with "hates"—the attitude that something will never work, an attitude he frequently saw directed at Allan Savory by his critics.

"I can't say enough about how much Allan Savory has created a new awareness and knowledge in the field of resource management," Kirk continued. "Much of the material I use in my role as an educator and consultant is based directly on his work in creating a holistic decision-making model. I consider it a great privilege to have worked so closely with him over the years. I was blessed to work with people who looked at the positive side of life and worked hard to get things done," he said. "And I still am."

I consider it *my* great privilege to have worked closely with Sid and Kirk over the years. When I cofounded The Quivira Coalition I chose another quote from Wendell Berry as our official motto: "You can't save the land apart from the people, to save either you must save both." I did so for two reasons: first, my background in anthropology and archaeology taught me that culture and the environment were mutually interdependent, historically and globally, and that no matter how far industrialization took us from our agrarian roots, we were *still* dependent on nature.

Second, I wanted to push back a bit against a major paradigm of the environmental movement, of which I was a member at the time, which said that nature and people (their work, specifically) needed to be kept as far apart as possible. Of most concern was a version of this paradigm that I heard voiced repeatedly by fellow activists: that environmental problems could be solved with environmental solutions largely devoid of culture or economics. Saving an endangered species, for example, meant either saving its habitat in a park or preserve or simply ejecting (without redressing) the destructive economic forces that imperiled the species in the first place. In other words, according to this line of thinking the natural world could be "saved" apart from an effort to "save" ourselves. Intuitively, I knew this argument was wrong.

But that was the problem. While I was happy to invoke the authority of Mr. Berry, whom I had just met, the truth was I didn't really know what his quote meant, not in detail. But that changed as I spent time with Sid on his ranch and Kirk in the workshops. Thanks to their mentorship I began to understand that "saving" means replenishing, rebuilding, renewing, reaffirming, restoring, and a hundred other acts of strengthening the bonds that tie people to the land and to one another. They helped me see that Wendell's assertion was more than just a quote—it was a possibility. And that gave me hope.

Chapter Three

GETTING INTO THE GAME

JT Cattle Company *Newkirk, New Mexico*

John Wayne must be rolling over in his grave.

This thought crossed my mind as I sat in the back row of a herding clinic taught by Guy Glosson, Tim McGaffic, and Steve Allen, ranchers and trainers who practice a type of low-stress livestock handling that emphasizes patience and kindness toward animals.

Stop the whooping and hollering when moving cattle, they instructed the twenty or so ranchers in the room. No more electric prods, sticks, or aggressive attitudes either. Throw away your conventional ideas of controlling animals by use of fear, pain, or other forms of stress-inducing pressure, they said.

"Consider not wearing sunglasses when approaching cattle," said Guy, the cherubic, bespectacled manager of an award-winning ranch in West Texas. "You're the predator and they're the prey, at least that's the way they look at it. If they can't see your eyes it may make them more nervous as they may not be able to judge your intentions."

I smiled to myself, imagining what the Duke would say to this. "Hell, you're *supposed* to make them nervous," the actor might growl. "That's the point. What is this, some sort of New Age Ranching?"

"If cattle get worried," continued Guy, "you've taken the first step toward losing control of the herd. Animals want to feel secure. But they won't feel secure if you're yelling at them all the time. Your job is to treat them with respect."

I could hear the Duke groan. Yelling at cattle and prodding them into action was as old as, well, the movies. Older, actually—Tim opened the class with a history lesson about how livestock have been manhandled in the West since the Civil War. Stressing cattle was part of ranching culture and is still standard practice on many ranches today. Perhaps that explains why Baxter Black, the most famous cowboy poet (and former large animal veterinarian) in America today, once challenged Guy over the idea of low-stress handling with a simple, steely: "*Why?*"

"I told him it is all about the health of the animal," Guy said. "Consistently handling animals without scaring them allows trust to be formed. This trust helps animals to remain calm and that equates into a healthier immune system and better response to vaccines and other medications they may need."

"I also told him that it was less stress on the handler too, which made us healthier," said Guy with his easy laugh. "But I don't think I convinced him."

But many others are convinced. That's because, as I learned that day in class, there are many reasons why low-stress handling is useful.

One reason is economics. The margin of profit for ranchers who sell cattle on the commodity market is literally counted in pennies per pound. The stress put on cattle as they move from the ranch to the feedlot and then to the slaughtering facility can "shrink" an animal's weight by 1–15 percent. That can be as much as $125 per animal, which is real money.

Stress can also make an animal more susceptible to disease, often requiring additional medicines and additional costs. According to Guy, stress can also affect pregnancy rates in cattle, the bread-and-butter of a rancher's bottom line. It all adds up quickly in dollars and cents.

Another reason to consider low-stress methods, Guy argues, is a philosophical one: we should treat animals with more respect than we do now. "Out of necessity we built a system of handling cattle based on fear and intimidation," said Guy. "The early herds of longhorn in Texas were wild animals, so the men who handled them had to be as tough and wild as the cattle. But that system is no longer needed."

"For grazing animals like cattle, the most dangerous predator on earth is a young human male," Guy continued, smiling again. "Until trust is established, animals will always perceive humans as a threat. And we don't want that. These animals are now domesticated and for the most part they depend on people for their every need. If we want them to perform at their best, they must not be afraid of the person caring for them."

Guy's argument fits with a changing value in American society—a rising intolerance for practices that are considered cruel to animals. As an example, the natural food chain Whole Foods recently entered into a partnership with the animal-rights group People for the Ethical Treatment of Animals to certify the welfare of animals that become the pork, beef, and poultry in their stores.

The Humane Society of the United States champions the techniques taught by Steve, Guy, and Tim. "Low-stress handling directly addresses some common animal welfare problems and is important as consumer concern for the welfare of farm animals increases," said Dr. Jennifer Lanier, director of scientific programs for the farm section of the Humane Society. "It is also inexpensive to implement and can immediately improve the lives of animals."

I imagined another grunt of disgust from Mr. Wayne.

According to Guy, everything starts with the predator-prey relationship and the effect such things as noise, size, distance, and motion have on cattle, which, like many animals, have well-defined zones in which particular actions trigger particular responses. For example, the *recognition zone* is where the animal takes notice of you and tries to determine your intent. The *flight zone*, when crossed, will cause the animal to move away from your approach. Violating this zone suddenly means you are likely to encounter an angry or panicked animal who has perceived you to be a threat.

The key to successful low-stress management of animals is something called "pressure and release." Your presence (as predator) creates pressure that an animal (as prey) wants to relieve. The critical moment happens when *you* choose to reduce the pressure instead of the animal doing it for you by fleeing. You do this by stepping into the animal's flight zone in such a way as to pressure it in a direction or manner you intend for it to move, and then *back off* when the pressure is no longer needed—before the animal runs off. Animals learn from the release of pressure, not the pressure itself.

The whole idea is to use a "law of nature" to positive effect.

For example, Guy teaches his students to approach an animal on foot in a nonthreatening manner, often zigzagging as the distance closes. When the animal sends a signal, such as raising its head, or widening its eyes, the student stops, or backs up a step or two. If an animal moves off, then the student is too close, or has done something threatening, and Guy says to start over.

"You're trying to start a conversation with the animal," said Guy. "You're not trying to tell him you're a nice guy or anything, because you're not. You're still the predator. Instead you're trying to communicate mutual respect. And you want to keep the conversation going as long as is necessary to get the job done. And you need to let the animal know when the conversation is over."

The modern concept of low-stress livestock management, as I learned from my visit with Steve Allen at his ranch, was developed by Bud Williams, a Canadian rancher who has spent his entire life studying how to handle animals respectfully, swiftly and easily, including reindeer, elk, sheep, and wild cattle. The key, Williams learned, was to pay attention to the instincts of the animal.

"We need people that are more sensitive to what the animal is asking us to do," said Williams recently. "If we would be more sensitive to that, then these jobs that we work on would be so much easier to do." According to Williams, it's all about communication, and not just between human and animal, but among people too. If you can't communicate your ideas of what you're doing, you probably also can't get it done.

"We always work at a level where we barely get it done," he continued. "We get as good as we need to get. We've reached a point now where we need to get better."

After lunch, Guy, Tim, and Steve led the ranchers outside where they would spend the next two days herding cattle around a large pen—on foot—learning to recognize the various zones and how to respond properly. It was a curious image, watching a rancher zigzagging slowly around the pen, head lowered slightly, stopping and starting, sometimes backing up as he or she tried to gently coax the small herd of cattle through an obstacle course of orange cones. It was even odder to hear the applause of the other students when the course was successfully navigated.

Clearly, this wasn't your grandfather's ranching anymore.

Sitting on the tall fence, I wondered: what sort of rancher is attracted to this stuff? What makes someone open to new ideas when another person, often a neighbor, will resist, refusing to look, learn, or listen? I know a rancher in eastern New Mexico, for example, who switched his management to a rotational system and watched springs come alive again with water. Grass grew tall and thick. His cost of production dropped in half and his profit tripled. But twenty years later, not a single neighbor had changed their management style despite his obvious success—and despite his willingness to teach what he had learned.

I glanced around at the students. I knew from introductions that morning there were two types of ranchers in the workshop: oldies and newbies. Most were oldies—folks from multi-generational ranch families—though many of them looked like they were members of the "next generation." Their motivation for attending, I suspected, was rather straightforward—they wanted to stay on the land and continue in a lifestyle that they loved. Why they were here specifically was probably a blend of curiosity, financial anxiety, and an innate desire to learn. The motivation of the newbies—those individuals just getting into the game—was harder to surmise, however. That's because it merged into a larger question: Why would someone go *into* ranching today?

Many weren't, of course. West-wide, more and more ranches are being purchased by nonranchers, often by wealthy individuals from urban centers. This is part of a larger trend in the region, a small but significant "reverse flow" of people moving back into the country as retirees, second-home owners, telecommuters, and other refugees from the harried existence of big cities. More often than not, they have the financial resources to shape country living to their liking. For example, in the 1980s and '90s, the media mogul Ted Turner bought huge ranches in Montana, New Mexico, and Nebraska, among other places, and replaced the cows with bison. He also implemented a costly endangered species recovery program. Neither activity endeared him to the ranching establishment, of course, but both indicated an emerging trend in land ownership and management in the West that steered clear of agriculture.

Others, however, were willing to take the plunge and become cattle ranchers. Why, I wondered? Not for the money. There's an old joke told by ranchers: how do you make a small fortune in ranching? Start with a big one. Thirty years ago it took the sale of eight steers to buy a new pickup truck; today it takes twenty-five. Or more. This is one of the principal reasons young people are not following their parents into the family business: it's almost impossible today to make a living producing beef full-time. Many ranchers, in fact, have part-time or full-time jobs in town now.

Newbies aren't getting into the livestock business for the social headaches either. Public lands in particular have their own set of challenges, not the least of which is intense scrutiny by environmentalists. But private lands ranching is getting tougher too, especially in these dry times when the pressure to graze more cows to pay bills or buy that pickup truck runs smack into Mother Nature. For private land owners, the brass ring of real estate development looms larger every day, especially in places like western Montana, where a square mile of land (640 acres) recently fetched $10 million! Too often the social and economic headaches, frequently combined with reduced interest in ranching by the next generation, cause the rancher to sell his or her property to a subdivider, who then resells the property as forty-acre ranchettes, with all the roads, pets, and power lines that come with them. As a result, acre by acre the West's vaunted wide open spaces are being busted into bits.

So why get into the game? I decided to ask Jim Thorpe, whom I met at the herding clinic that day. He and his wife Carol, a middle-aged couple from Santa Fe with no background in agriculture, decided to take the plunge and become cattle ranchers. Boyish, and sporting an impish sense of humor, Jim responded with a big smile to my question. He admitted that the romantic image of ranching played a role in their decision, but then he surprised me. He said the main reason was that it gave him a chance to exercise his passion for ecology.

"Ecology?" I asked, expecting another groan from the Duke at any moment.

"That's right," Jim replied. "Ranchers have become applied ecologists. Most don't look at themselves that way, and most don't like the word 'ecology' very much, but nowadays it's all about stewardship, not just food and fiber."

"But it's also about having a passion about what you are doing," he continued, loquaciously. "Like many, I'm grateful for the privilege of being allowed to take care of a little piece of the planet for whatever short amount of time we have."

Curious about both Jim's theory of ranchers as applied ecologists and their decision to get into the game, I accepted his invitation to visit the ranch, which is located north of Santa Rosa, in east-central New Mexico, nestled in the hilly transition zone between mountains and plains. Despite its remoteness, Jim and Carol selected it after a long search because, as Jim put it, they "wanted a place in the middle of nowhere, and a couple of hours from everywhere."

The Thorpes' adventure began in 1998 when his family sold Bishop's Lodge, a historic Santa Fe landmark. After two decades in the hotel business, Jim and Carol suddenly found themselves out of work. Casting about, and with two children preparing to leave the nest, they decided to buy a ranch.

"I had friends in ranching," explained Jim, "which meant I had just enough knowledge to be a little dangerous. I knew, for instance, from a classical economics perspective, ranching wasn't the most rational way to invest an unexpected windfall. But then humans aren't terribly rational."

Before taking the plunge, however, Jim did some homework. Over the span of two years, he visited ranches he knew, read a pile of books and articles, went to meetings and workshops, scoured the Internet, and made use of the business skills he had acquired years ago in an "MBA light" course at the University of New Mexico. He eventually joined the Society for Range Management, the New Mexico Cattlegrowers' Association, and The Quivira Coalition.

Much of what he sought to learn was ecological, a term that struck a chord with Jim in ninth grade and resonated all the way through college, including his participation in the original Earth Day celebration in 1970. "I soaked up the ideals of the environmental movement," he recalled, "but I never thought that I might end up utilizing it to earn a living."

As we talked in the living room of the ranch house, I noticed evidence of Jim's continuing quest for knowledge, as well as his liberal arts background. On the table sat a book of Shakespearean plays, a collection of essays on wilderness (heavily underlined), and a large, colorful tome on the sins of cattle ranching in the West. I asked about the last publication, whose author was well known for his position that ranching was an "irredeemable" occupation. Jim smiled and said it was good to learn what the other side was thinking. He went on to say that the anti-cow crowd had important points to make, as do the traditional ranchers who are his neighbors. It's never either/or, he insisted.

What about the Shakespeare? I asked. It was good to keep up with the classics, he replied. And sometimes he found it useful in unexpected ways.

"I sometimes like to ask myself, 'What would Socrates do on the ranch?'" Jim said, smiling impishly again.

Despite their preparation, there was nothing quite like the crash course of moving to the twelve-thousand-acre ranch itself. "All cows looked alike and all grass looked alike when I started," said Jim, "which may be why a group of vultures eyed us suspiciously our first night on the ranch. Maybe they thought we looked like easy pickings."

To avoid the vultures, the Thorpes sought out "old knowledge" from traditional producers and agricultural specialists. For example, they enrolled the ranch in Texas A&M's Cooperative Extension program called Standardized Performance Analysis, which provides data on cow-calf operations and gives participating ranchers an opportunity to compare their operation with others of similar size and geographic range.

At the same time, the Thorpes extended their quest for knowledge to the "new school" of ranching. They read Allan Savory's treatise, *Holistic Management*, for example, and hired Kirk Gadzia as a consultant to their ranch. They subscribed to the monthly alternative newspaper *Stockman Grass Farmer* and read books by its publisher, Allan Nation, including *Knowledge Rich Ranching* and *If You Want To Be a Cowboy, Get a Job,* which challenged traditional paradigms about profitability and management in the livestock industry.

Jim saw the blending of the old and new as a sign of the changing times.

"In the Old Ranch, the focus is on the cattle, including how to increase income from beef," said Jim. "In the New Ranch, the focus is on the grass and soil, and on how to diversify the business. Right now I'm trying to coordinate the two. I'm interested in ecological services, such as increasing wildlife habitat and improving water quality, for example, but right now I've got to focus on beef in order to pay the bills."

Fortunately for the Thorpes, the blend of old and new is paying off, literally. Although the ranch has the capacity to feed at least 300 bovines, the Thorpes chose to start with 200. Over time, through careful management, they raised the total to 250 head of cattle, all in the middle of an extended period of drought. The income from the cattle is covering their expenses and paying the interest on their land debt. Although they aren't seeing the sort of profit they saw in the hotel business, they have been struck by other parallels with their work in that business.

"Carol's passion is for the animals," Jim said with a smile, "so she's in charge of personnel. I'm the facility manager."

Clearly enjoying themselves, the Thorpes keep moving ahead. They participate in a New Mexico State Land Office program that reduces grazing fees 25 percent in return for good management and monitoring. They are enrolled in New Mexico State University's Ranch to Rail program, which provides carcass information and genetic data so they can improve the quality of their beef. They are also exploring the idea of creating "custom habitat" for wildlife, which means managing portions of the ranch for the needs of particular species, especially those that teeter on the edge of survival.

"I feel like sometimes we're bridging two worlds," Jim said. "I'm trying to provide values that society wants while placing a traditional cattle operation over it at the same time. It's tricky, but I think it's going to work."

Still, he admits that the challenges are daunting. A globalizing economy (such as beef from Brazil), political marginalization, antagonistic attitudes from environmental activists, indifference from urbanites, even the test of rural isolation can ruin the spirit of many first-time ranchers fast.

Jim's response is to remain philosophical.

"Everyone wants a sense of purpose, whether they know it outright or not," he mused. "We must deal with a tragic sense of existence, which compels us to leave the world in as good or better condition than when we got it."

"It's all about our moment in the sun, literally," he continued, "and our relationship to nature. Ranching is a great combination of mental and physical; throw in a few philosophical paradoxes and it really gets interesting."

Socrates, in chaps.

Driving away, I mused over Jim's point about paradoxes. A few had been on my mind for a while. Why, for instance, was herding considered a new, even radical, idea among ranchers when it's been going on for thousands of years around the planet (where it is sometimes called transhumance)? Herding is as old as agriculture, yet it somehow became buried in the avalanche of change brought on by industrialism and the barbed wire fence. And despite its many advantages, herding is largely dismissed by ranchers as untraditional, even if its adoption meant that they could keep living the lifestyle that they considered essential to their sense of self-worth.

Which led me to another paradox: why do traditional ranchers call themselves the "original environmentalists," as one prominent rancher did in a letter he sent me critical of our work, when one could plainly see overgrazed land all across the state? Of course, in a sense he was right:

ranchers *are* sensitive to the environment, or else they would be out of business sooner or later. But I had real trouble squaring that attitude with the poor condition of many parcels of land that I (and others) saw under their care. Sure, conditions on the ground had improved from the cattle boom years at the turn of the previous century, when serious ecological trauma occurred as a result of massive overgrazing by livestock, but that hardly excused poor management today.

Which led to another conundrum: why did environmentalists persist in their efforts to end livestock grazing on public lands (as some still do today) despite numerous examples of good stewardship by ranchers in a variety of landscapes? And why do they do so against the backdrop of rapid fragmentation of private land in the rural West? Eliminating livestock production on public land will likely result in the sale of the rancher's private land, probably to developers, since many operations need both in order to remain economically viable. And what about stewardship? Restoration? Local food production? Did environmentalists expect the federal government to do these things?

Which leads yet further: if the feds can't or won't do these things (and if environmentalists won't either), then how do we get the best parts of a private land ranch operation—its innovation, flexibility, efficiency, and entrepreneurial spirit—placed within the framework of public lands? Or should we even try?

On the highway, headed home to Santa Fe, I thought again about Jim's point about ranchers being applied ecologists. Perhaps the answer—or at least a rebuttal—to these paradoxes lay in the realm of land health. Unquestionably, there is a great deal of solid ecological knowledge out there among environmentally minded professionals; likewise, there are many generations' worth of land management knowledge among ranchers. Fusing the two somehow had to be the key to progress. This was why the New Ranchers appealed to me—they bridged the gap between theory and practice in a way that benefited both land and people.

There was another reason: in the twenty-first century, with all its uncertainty about climate change, energy depletion, potential food and water shortages, declining environmental health, and possible political and social unrest (due to one or more of the above), we will all be "getting into the game" to one degree or another whether we want to or not. What was familiar and comfortable may soon no long be either. The only thing we know for sure is that things will change, perhaps drastically. We may be forced into the game. Which leads to one more thought: if it is all right for ranchers to become applied ecologists—conservationists, essentially—then shouldn't it be equally all right for conservationists to become ranchers, or at least local food producers?

Actually, that's not such a bad idea. Maybe even the Duke would approve.

RANCHING
WITH WILDLIFE

Sun Ranch *Upper Madison Valley, Montana*

When Todd Graham took over as manager of the Sun Ranch, located in Montana's wildlife-rich Madison Valley, northwest of Yellowstone National Park, he didn't realize that camping with cattle was part of the job description. But that's exactly where the bookish thirty-four-year-old found himself less than a month after starting work—sleeping in a tent in a pasture amidst a herd of yearling cattle. And he was there for a good reason: he had a den of gray wolves as neighbors.

"Since the den was only half a mile away, the chances of action were high," recalled Todd, who had decided to place himself between the den and the cattle. "I crawled into my sleeping bag and inventoried my gear: bear spray, twelve-gauge shotgun loaded with rubber bullets, two monster flashlights capable of lighting up the mountain, hunting knife, and running shoes for sprinting."

After being hunted almost to extinction in the lower forty-eight states, the gray wolf was formally reintroduced to Yellowstone and central Idaho in the mid-1990s by the Clinton administration, over vehement objections from the livestock industry and its allies. Among a litany of complaints, ranchers feared for the safety of their animals, particularly calves, as well as for their own individual well-being among these wild creatures. Over the next decade the population of wolves expanded rapidly (to over 1,500) and were soon heard howling on private ranchland throughout the greater Yellowstone ecosystem. Inevitably, some cattle were attacked, and in response some wolves were shot by government agents, upsetting environmentalists and other wolf advocates. Tensions ran high on all sides.

But not on the twenty-five-thousand-acre Sun Ranch, whose new owners didn't view wolves as a threat but as an important part of the ecosystem. Strategically located in a Montana valley that is home to a wide variety of wildlife, including grizzly bears, wolverines, mountain lions, mountain sheep, and mountain goats, the Sun Ranch is the winter home to four thousand head of elk. The ranch is also located in spectacularly beautiful country, which, combined with the wildlife appeal, is why so much of the upper Madison Valley, the portion closest to Yellowstone, has been purchased by wealthy, out-of-state admirers. What made Todd's employers different, however, was their intention to do more than simply "ranch the view."

Boasting one of the oldest brands in Montana, the property was called the "Rising Sun Ranch" until the outbreak of World War II. Its previous owner was a Hollywood star who damaged the land, according to Todd, though not deliberately, by overgrazing it with livestock in an effort to maximize profits. When the ranch changed hands in 1998, the new owners had a different mission in mind: to find a way that wildlife—elk in particular—could coexist with livestock.

By the time Todd took over, however, the owners had set aside one-third of the ranch as an elk reserve, where cattle were off-limits, under the prevailing assumption, common among many conservationists, that wildlife and cattle need to be segregated. The trouble was the elk had stopped using the pasture too. The grass was old and rank, recalled Todd, and the elk didn't like it. Todd received permission to try a different approach.

He decided to "freshen up" the forage with cattle during the growing season, using them as a conservation tool. Employing single-strand polywire (high-quality plastic) electric fencing and portable posts, Todd created a sequence of five-hundred acre pastures in the elk reserve and then turned out a portion of the ranch's 1,300-head cattle herd for six to eight days in each pasture early in the summer. The cattle trampled and ate the old grass along with the new.

"Our goal was to increase the quality and quantity of forage for the elk by making the cattle disturb the old, unused forage," said Todd. "The results were outstanding. We grew a great deal of new grass and the elk returned in big numbers."

And when the grazing rotation was completed, the polywire fence was rolled up and removed, leaving not a trace. "Polywire is an amazing tool," said Todd. "In addition to being just right for planned grazing, elk can see it easily and will jump over it; deer too. Antelope go under it."

Not many ranchers use polywire, however, just like not many are willing to camp out with wolves. Most ranch the old-fashioned way, with lots of barbed wire, and most still see predators, including coyotes, as nothing but trouble, which demonstrates how far someone like Todd has come.

Born and raised in Big Piney, Wyoming, where his parents were schoolteachers, Todd worked on a nearby ranch that employed progressive ranching methods. Intrigued, he pursued these ideas at the University of Wyoming, where he studied range science and enrolled in a Holistic Resource Management class taught by Kirk Gadzia.

"All of a sudden a door came open—it was like four years of college in three days," he recalled.

Emboldened, Todd embarked on a successful career of ranch management and range monitoring as a private businessman. But ranching with wolves was another matter. For three years Todd had consulted to the owners of the Sun Ranch before becoming manager, so he knew that he might face the challenge of running livestock in the presence of wolves. But so soon?

"After zipping up the bag, turning off the headlamp, and settling in, two thoughts raced through my mind," he said. "First, I have no idea what I'm doing. Second, there's no way I can pull this off alone. The next day I asked for help and began learning the power of collaboration not only in dealing with wolves, but in managing a landscape."

He called friends and acquaintances in the conservation community and before long he had a steady stream of volunteers camping with him on the ranch, each tasked with the job of observing the wolf den and discouraging the animals, if necessary, from approaching the cattle herd. It worked—not a single bovine fell prey to the carnivore.

At the same time, Todd became involved in the Madison Valley Ranchlands Group, a rancher-led collaborative organization whose mission is to enhance the economic viability of family ranches as well as to maintain healthy grasslands and wildlife habitat. Before long the group gave Todd an unusual chore: to write a grazing plan for the entire watershed, which meant incorporating the private land of the sixty or so ranch families that lived and worked in the northern half of the valley with that of the out-of-state owners that dominated the southern half *and* the state and federal land that ringed both sides of the valley. Could one herd of cattle, for instance, be driven from one end of the valley to the other and back again in one grazing season? It was an intriguing and audacious concept that Todd eagerly tackled, even though he knew that it would take many cups of coffee in many different kitchens to get such an idea off the ground.

"I wouldn't have said yes," said Todd, "if I didn't think it was possible."

Inspired by the tangible benefits of building bridges between diverse interests, Todd seized another opportunity in June 2003 when he accepted an invitation to join the board of directors of the Greater Yellowstone Coalition (GYC), a twenty-one-year-old conservation organization with thirteen thousand members and a lengthy history of aggressive advocacy, including litigation, in the Yellowstone region. In 2006, he became chair of the board. He was asked to join, he believes, because GYC was beginning to see that cooperation was as important to the success of its mission as confrontation. Todd also saw his involvement as an opportunity to help environmentalists learn about progressive ranch management. But he isn't opposed to the organization pushing back against poor cattle management.

"On one hand, the organization devotes resources and skilled staff to helping ranchers and communities adapt to changing times and challenges," he said. "On the other, GYC gets aggressive where ranchers and the agencies are demonstrably doing a poor job of managing the public's lands. I like both approaches."

But can wolves coexist with humans and cattle in a busy valley such as the Madison, divided between intense recreational activity and second-home ownership in the south and agricultural use in the north? Todd thinks they can, but the keys are communication and innovative management. He believes that if you ranch to improve wildlife habitat and work collaboratively with friends and neighbors, as they are doing on the Sun, your chances of coexisting with wolves dramatically improve.

"Wolves are yet another challenge facing ranchers today," said Todd. "Succeeding in their presence will require us to work together. We must learn as much as we can about their behavior and adapt our practices with new knowledge. Today, our teachers may be predator lovers and wildlife biologists, rather than other ranchers and universities."

Then he added, with a smile, "Welcome to the New Ranch."

Chapter Four

OUT OF
COWBOY ISLAND

The Former Vail & Vickers Ranch *Santa Rosa Island, California*

Nita Vail's story, like any good western, is part fairy tale, part tragedy, with a moral and a happy ending to boot.

I heard the details when I crossed paths with Nita at a conference at a former ranch turned resort on the outskirts of Tucson, Arizona, where we joined two dozen other citizens from around the region to talk about the "ranching crisis" in the West. That's what we called it anyway; in reality it was a "sprawl crisis," because our concern focused on the rapid rate at which former ranch lands were being converted to housing developments. For some in attendance, this crisis meant the loss of ecologically significant open space to suburban and exurban growth. For some, it meant the loss of critical food production capacity, as farms and ranches disappeared under the bulldozer's blade. For others, it meant the loss of a culturally important lifestyle, as cows were relentlessly replaced by condos. For all, it meant the loss of much of what we loved about the rural West—its land and people.

It isn't just a western crisis either. According to the nonprofit American Farmland Trust, every minute of every day the nation loses two acres of agricultural land to development. And the rate of loss is speeding up:

the nation lost farm and ranch land 50 percent faster in the 1990s than in the 1980s, with much of the loss taking place on prime agricultural land. For example, in the central valley of California, which annually produces more than $10 billion worth of food products, almost one hundred thousand acres of fertile farm and ranch lands were paved over in the 1990s. Nationwide, over the past twenty years the amount of acres per person consumed for new housing has doubled; and since 1994, housing lots of ten acres or more have accounted for 55 percent of all land developed. The deleterious consequences of all this growth extend beyond food production. Farms and ranches provide wildlife habitat, they help protect watersheds, maintain air quality, and provide scenic and recreational opportunities—all lost when we begin pouring concrete and laying asphalt. It is a crisis, in other words, that is felt by all Americans.

My role in the conference was to talk about The Quivira Coalition's efforts to build bridges between ranchers and conservationists around models of progressive stewardship—models that often provided an economic boost to the landowners. I argued that if a conservation organization wanted to protect the biological integrity of a farm or ranch, it was far cheaper and more effective to help the rancher stay in business than to buy his or her ranch when it was put up for sale.

Nita was there to talk about a different strategy to keep ranchers on the land. And because this is the happy ending to her saga, I'll start there.

As the executive director of the California Rangeland Trust (CRT), it is Nita's job to protect as many private ranches as possible using an important tool called a conservation easement, which is a legal mechanism by which a landowner can strip his or her land of its development (subdivision) potential while maintaining control of the property. Often easements are donated or sold to a land trust, which holds them in perpetuity. In the United States, the land trust movement took off in the mid-1970s and today protects more than 5 million acres nationwide from development.

Fair-skinned, blue-eyed, and as composed in appearance as she is in manner, Nita looked more like a successful businesswoman than the daughter of a multigenerational ranching family. I was sure she was equally at ease working her Blackberry wireless as working cattle from horseback. Her career included an MBA in agribusiness from the University of Santa Clara and an appointment by Governor Pete Wilson to serve as the assistant secretary of agriculture and environmental policy in California's Department of Food and Agriculture.

In 2001, she became executive director of the CRT. Of approximately two hundred land trusts in California, the CRT, headquartered in Sacramento, is the only one focused exclusively on ranches and rangeland conservation. Under her energetic and determined direction, ably assisted by a board of directors composed of ranchers, the organization has become one of the largest land trusts in the state. At the time of our conference it had 173,000 acres under conservation easements statewide, and more than 500,000 acres pending in application—a remarkable achievement for a land trust that had been in existence only since 1998.

"We were lucky," she said. "We got our first easement right out of the chute. That gave us credibility and created trust, which was important given the general attitude towards conservation easements at the time."

That's because until the mid-1990s, easements were almost exclusively a tool employed either by large conservation organizations such as the Nature Conservancy, small, urban-based land trusts, or governmental entities. Ranchers were suspicious of all three for a variety of reasons. Moreover, many ranchers were staunch defenders of their private property rights and distrustful of any organization or agency that proposed to strip out any part of those rights, even if doing so helped keep the family ranch in business.

The atmosphere changed dramatically in 1993 when the Colorado Cattlemen's Association decided, over some noisy dissent among its members, to form its own land trust. Their idea was simple: create a land trust for ranchers controlled by ranchers. Membership was strictly voluntary. No one was required to sell an easement to the land trust, but if he or she did then they were assured that it would be protected in perpetuity as a working ranch. The idea caught fire. Before long, the success of the Colorado Cattlemen's Agricultural Land Trust inspired their friends in California to give it a try.

"Times were tough," recalled Nita. "Cattle markets were poor and pressures to develop were building on all sides. Also, generational succession is one of the great challenges facing California's ranchers. We saw what was happening in Colorado and decided it was time to take charge of our own destiny."

According to Nita, there are multiple financial benefits to easements. The landowner may be entitled to a charitable tax deduction; an easement can lower the taxable value of the land for estate tax purposes; a person inheriting the property may be eligible for estate tax benefits; and a conservation easement may lower property taxes.

All easements do is restrict development rights in perpetuity to agricultural and open space uses (such as recreation), nothing else, she continued. Contrary to what many ranchers believe, easement agreements should not interfere with the day-to-day work of the ranch. The property owner still holds title to the land, can limit access, even if the easement was purchased with taxpayer money, and may sell, donate, or transfer the property as he or she sees fit.

Legalities aside, said Nita, the principal value of conservation easements is simple: it keeps families on the land.

"Open rangeland is best protected by the ranchers who make their living from it," she said. "An easement allows the landowner to receive compensation for the open space values his or her property provides but still maintains it as a working landscape." Easements cost less than purchasing the land outright on the open market at subdivision rates. They're cheaper to maintain too because the landowner takes care of the property, and the property remains on the county tax rolls. Of all the reasons to do a conservation easement, however, it is the desire to keep the land whole and in agriculture that motivates most ranchers, she said.

"Easements aren't for everyone," she cautioned. "Some of the language in the agreements is draconian, so you've got to be careful. And some of the criticism of easements is legitimate, but people forget that doing an easement is a property right too."

Nita pointed out that ranch families own or manage 22 million acres of land in the Golden State. By 2040, it is estimated that the state's population will swell from its current 32 million to more than 50 million people. That means more pressure to build subdivisions, strip malls, and freeways on or through private land. It also means more temptation for ranchers and farmers to sell out. And once land becomes fragmented, all the king's horses and all the king's men won't be able to put Humpty Dumpty back together again.

Nita cited two more reasons why protecting California's ranch lands has ecological benefits: virtually all water consumed in California flows over rangelands at some point; and 95 percent of all threatened and endangered species in California are found to one degree or another on private ranch lands.

Despite these benefits, however, rancher-owned easements remain a contentious issue with some environmental organizations, who liken it to the fox guarding the henhouse. "Our organization is put under a microscope probably more so than organizations that have strictly an environmental purpose," she said, "when in fact we all want to do the same thing—protect the landscape. We are just trying to do it in a way that also makes economic sense."

It made sense to me. Still, I knew from bits and pieces of hearsay that there was more to Nita's story than the success of the California Rangeland Trust. I understood that she and her family had endured a serious heartbreak with their ranch on Santa Rosa Island, located northwest of Santa Barbara, in southern California. In fact, I suspected that her drive to save ranching in the Golden State was motivated by events on the island during her youth—events, I learned, that lent a melancholy tone to Nita's fairy tale.

At the conclusion of the day's proceedings, Nita and I retired to the cool shade of a table awning, sipping drinks as the heat of an early summer day mellowed into a lovely stillness. Once upon a time, the Vail family owned Santa Rosa Island, part of the Channel Islands group. Nita's great grandfather, part of a ranching family with roots reaching to southern Arizona, purchased the fifty-four-thousand-acre island in 1901. The rhythms of ranch life remained unchanged until 1979, when Congress created legislation expanding Channel Islands National Park to include Santa Rosa Island. The bill had the support of the Carter administration and was pushed by many environmental groups who said they were concerned about proper protection for rare plants and animals on the island.

The Vail family responded by trying to get their island exempted from the legislation, arguing that their progressive management worked in harmony with nature, and that the "no trespassing" restriction of private property provided solid protection to the whole island. They also insisted that commercial or residential development on the island was not part of their long-range plans. These messages fell on deaf ears. The family could not find political allies outside of the ranching community.

"The question that we asked but nobody answered was: protection from what?" Nita told me. "It felt like, from us."

Nita said a painful irony was at work: although the Vail family's good stewardship of the island's natural resources was one of the acknowledged reasons why Congress and environmentalists wanted the island "protected" permanently in the first place, the presence of their cattle was deemed "unnatural" by the same advocates, thereby requiring eventual removal.

"It hurt," she said, "because it was unfair. We were holistic managers before it had a name."

Nita noted that her family had few choices. Conservation easements, for instance, were not a practical option at the time. They were also swimming upstream against an environmental movement on a roll since Earth Day. For many activists, there was little or no gray area between preserving a place as a park and "losing it" to agricultural use. For a biologically rich landscape such as Santa Rosa Island, the choice was clear to them.

The Vail family struggled against park proponents until 1986, when they gave in to the inevitable and sold the ranch to the federal government, brokering a deal that allowed them to stay on the island, and in business, until 2011.

The story wasn't over, however.

A few years later, the deal was jeopardized when National Park Service biologists inventoried the ranch and discovered that several species of plants and animals were in danger of extinction, including the island fox, which eventually wound up on the federal list of endangered species. Park advocates, including biologists with other agencies, began to insist that livestock production had to be curtailed or ended, pronto. A series of complicated and disheartening—to Nita and her family—maneuvers followed involving various state and federal wildlife agencies as well as (behind the scenes, according to Nita) some key environmental organizations. Pressure began to build to break the deal with the Vails.

When the family continued to insist that a deal was a deal, an environmental organization sued the government to get the cows off, and won. The last cattle drive, memorialized in Gretel Ehrlich's book *Cowboy Island*, took place in 1998.

Years later, the outcome still rankled Nita.

"The 1916 Organic Act of the National Park Service contains an inherent contradiction which was played out on our island," she said. "Going back to pre-European landscapes and increasing public access seem incompatible."

She was referring to the National Park Service's mission, coded in its 1916 enabling act, to "conserve the scenery and the natural and historic objects and the wild life therein . . . by such means as will leave them unimpaired for the enjoyment of future generations."

"Not enough credit is given today for good stewardship," she continued. "The Park Service is not holistic. The ecological condition of the island is awful now, but no one talks about the grasses; all they can do is focus on the fox, which is being eaten by eagles."

In an effort to return the island to a desired "presettlement" condition, the Park Service removed all the feral pigs. With its prey base suddenly gone, the island's golden eagle population turned its predatory attention to the island fox (apparently the birds hadn't read the park's management plan), causing a crisis.

Nita isn't the only player in this tale that feels her family was treated unfairly. In October 2006, former Channel Islands National Park superintendent Tim Setnicka authored a three-part opinion piece in the *Santa Barbara News-Press* in which he wrote, "To this day, no one has shown that the ranching operation has permanently, significantly or irreparably destroyed park resources."

He went on to criticize in detail the behavior of federal and state agencies, as well as environmental groups, in their quest to push the Vail family off the island early. "In this process," he concluded, "the National Park Service has lost much of its credibility in the public eye about how and what it does to carry out its mission. In turn, the environmental community argues that it had to 'save' Santa Rosa Island resources, but the facts are that each year the island's resources were in better and better shape. They didn't need saving from the Vails' activities."

I suspect Nita took little cheer from the former superintendent's words. Still, the whole experience proved to be a motivating lesson for Nita, as she admitted to me. And with the California Rangeland Trust, she found her happy ending.

So, what is the moral to this tale?

For Nita, it was this: "We need to remember where we have come from and go forward in a positive way," she said. "We can't stay stuck in the past."

Our conversation came to an end when Nita excused herself to take a call on her cell phone. I ordered another drink. It was truly a lovely evening, reminding me of how much I missed the Sonoran desert of my youth. I grew up in Scottsdale, a tony suburb of Phoenix, and thinking about the day's events I recalled the proliferation of real estate signs along what was then the desert edge of my hometown. Many of the signs, including one titled "Cowboy Land & Cattle Company," had been defaced by anonymous vandals with a plaintive, spray-painted cry: "SAVE OUR DESERT." I remember asking my father what the message meant, even though I had a good idea.

Of course, the desert wasn't saved. In 1994, while passing through Scottsdale on my way home, I took a detour and searched for a small, ramshackle horse stable my father had rented called Powderhorn. In my memory, it sat far out in the desert, like a mirage, its only neighbor a funky palm tree nursery. Together, they formed an odd oasis in a sea of warm sand and creosote bushes. Searching, eventually I found what I was looking for—a generic sign on a generic wall in a generic sea of houses announcing "Powderhorn Estates." Gone to subdivision, every acre.

No one saved my desert. It's not likely that the desert around Tucson can be saved either. Nita may not be able to save ranching in California. The National Park Service may not be able to save the island fox. No one knows what my fellow environmentalists were trying to save when they pushed the Vails off Santa Rosa Island early. I wondered: can anything be truly "saved" in a world of relentless change? Can any species be fully "protected" in an age that looks to be dominated by global climate change, rising population pressures, and persistent pollution? Should we even try? Or should we try a different approach?

I took another sip of my drink.

The moral of Nita's story was different for me. It warned against the arrogance of certainty in a world characterized by increasing uncertainty. The Vails were punished for what looks like an erroneous belief that their activities were endangering the island's natural bounty. Ignorance and arrogance, in other words, were employed to "save" something that perhaps didn't need "saving."

Is Santa Rosa Island better off today without cattle? I can't say. Would the Vail family have done a better job of managing the island fox than the National Park Service? We'll never know. Does the island's status as a park guarantee that its biological resources will be "unimpaired" for future generations? No one knows. But did it have to be either/or? Park vs. ranch? Saved vs. unsaved? Or could a cooperative arrangement have been created that allowed the Vail family to continue ranching sustainably while government specialists worked to ensure the island's natural bounty? Not at the time. But perhaps things could be different in the future.

Chapter Five

CHANGING
THE WORLD

The Upper Eagle Creek Watershed Association
North of Morenci, Arizona

It is a sign of the changing times that a meeting in a one-room schoolhouse in a remote valley in the Blue Mountains of eastern Arizona featured a PowerPoint presentation. Less emblematic, but more important perhaps, was who did the presenting: members of the local ranching community. This fact didn't go unnoticed by the new supervisor of the Apache-Sitgreaves National Forest, who was visibly impressed not only by the presence of the technology in so remote a location, but also by the show itself.

She wasn't the only one impressed that day in 2004. As the show unfolded, detailing the group's ambitious goals and plans—all to be accomplished in concert with the U.S. Forest Service—it became clear to me that the eight families of the 220,000-acre upper Eagle Creek watershed had a good chance of succeeding. Why? Because they had decided to stop fighting the future. They were confronting changes and challenges with creativity, turning adversity into opportunity.

I'll admit, however, to some initial skepticism. When Frank Hayes, the Clifton District ranger for the Forest Service (and a Quivira Coalition board member), called me a few weeks earlier to suggest that I should check out what was happening on Eagle Creek, I wondered: was it worth the very long drive? Eagle Creek, after all, was just over the state line from Catron County, New Mexico, which had earned an infamous reputation in the 1990s as a hotbed of antifederal, antienvironmental activism. Furthermore, I knew that the Mexican wolf, exterminated decades ago at the insistence of the livestock industry, had recently been reintroduced only a few miles from the headwaters of Eagle Creek. I suspected it remained an unpopular decision among the residents.

I harbored doubts, therefore, that the area was suddenly on the front lines of the collaborative, watershed-based movement that had sprung up around the West in recent years. Still, I trusted Frank; and besides, I knew some of the Eagle Creek ranchers from workshops we had organized. And if Frank's boss was going to be there . . . OK, I told him, I'll make it. When my son, Sterling, who was five at the time, expressed a desire to tag along (his twin sister, Olivia, was sick), it became a family affair.

Driving southwest from Santa Fe, we sliced through contentious Catron County, stopping in Reserve, its biggest town, for a meal. In the mid-1990s, the three-person Catron County Commission made news when it passed an ordinance requiring residents to carry a sidearm. They also passed a resolution declaring primacy over federal lands— meaning that they believed their laws and regulations trumped federal jurisdiction on the national forests in the county, prompting other rural counties across the West to follow their lead. Called the "county movement" by observers, it was the latest iteration of the so-called Sagebrush Rebellion—an effort ignited in late 1970s by ranchers, loggers, miners, and others in an attempt to push back against new environmental laws (and shifting political winds) by making "states' rights" claims over federal land. The U.S. Constitution, they argued, was vague on the primacy of the federal government on territory outside the original thirteen colonies.

For a while, tiny Reserve was ground zero in this latest rebellion. Looking around that day at the town's few shops and fewer visitors after our meal, it was easy to get the sense that residents had a legitimate economic grievance. Half of the West is owned by the federal government, by one agency or another, and in some states, such as Nevada, the total rises well above 80 percent. Unquestionably, the restrictions, regulations, red tape, and intense scrutiny that came with trying to wrest a living from federal land made life increasingly difficult for ordinary rural westerners, fueling emotions that gave rise to the anti-federal backlash at the time.

Not surprisingly, the Sagebrush Rebellion cooled off with Ronald Reagan's election to the presidency in 1980 and the subsequent selection of James Watt, a natural resources lawyer from Colorado who sympathized with the rebels, to be Interior secretary. Order was restored, in a fashion—but only until 1993. That's when Bill Clinton moved into the White House, swinging the pendulum, once more, completely to the other side. Environmentalists now had the upper hand in Washington, D.C., and one of their first targets in the West was "grazing reform," which included an attempt to increase the fee ranchers paid the government to graze their cattle on public land. Politically, it was a huge mistake. The ranching community reacted swiftly, objecting both in court and in the court of public opinion, grinding most of the reforms to a halt. In the process, hot rhetoric from both sides of the grazing debate fanned the fires of the Sagebrush Rebellion back to life, to the dismay of many of us watching from the sidelines.

Sterling and I put Reserve behind us. After crossing the Arizona state line, the road suddenly dropped spectacularly and before I knew it we had descended through two or three ecological life zones, leveling out in a familiar landscape of muted desert browns and pale greens. It was new land to Sterling, however, and he peppered me with questions about our mysterious destination.

We turned right at a highway intersection and headed north toward the colorful country surrounding Morenci. Before long we were gawking at the awesomeness of a huge open-pit copper mine that flanked both sides of the little road. Dump trucks the size of small buildings rumbled below us. Sterling insisted that we stop at an overlook. As he peered into the depths of the monstrous mine, I thought about the toy dump truck he had back home, prompting a philosophical observation:

the human ability to move mountains when we wanted to, literally, is as innate as it is incredible to behold.

The tenor of the Clinton years was symbolized by an event that occurred in Mineral County, central Nevada, when an angry county commissioner named Dick Carver jumped on a bulldozer and reopened a road on public land in defiance of the federal government. In his shirt pocket was a copy of the U.S. Constitution. For his effort, he made the cover of *Time* magazine in 1995 under the headline "The War on the West"—meaning the Clinton administration's supposed war on ranchers, loggers, and miners. The press loved it. It was high noon—David versus Goliath, all over again.

Goliath won. It wasn't even close: the courts ruled repeatedly against the counties and their allies, saying that Dick Carver's beloved Constitution was clear: the federal government controlled federal land. Period. It would prove to be the last hurrah for the Sagebrush Rebellion. The county movement faded away. One of its principal advocates, a noisy property-rights group called "People for the West," closed its doors. Anger and frustration among rural residents were gradually replaced by a general feeling of resignation, including in places like Eagle Creek, our mysterious destination.

Twisting through the mountains above the copper mine, Sterling and I eventually turned west, off the narrow (and stomach-churning) highway onto a well-maintained road that steadily lowered itself to the grade of Eagle Creek, visible in the distance as a ribbon of leafy trees among the grasslands.

Some places in the remote West, however, didn't mourn the end of the Sagebrush Rebellion. That's because they chose a different path, preferring compromise and common ground to confrontation. For example, down in the southeastern corner of Arizona, along the Mexican border, a group of ranch families linked arms in the mid-1990s with allies in the Nature Conservancy and friends in the state and federal agencies in an effort to protect a million-acre watershed from ruinous development. In doing so, they sparked a counter-movement to the hard-headedness that had characterized the "old schools" of environmentalism and ranching that had dominated the political landscape for so long. One of the group's leaders, rancher Bill McDonald, dubbed this movement the "radical center."

In 2003, Bill and I were members of a group of twenty ranchers, environmentalists, and scientists that met for forty-eight hours in an Albuquerque hotel in an attempt to take back the American West from the years of pendulum-swinging and divisiveness that we thought were jeopardizing so much of what we all valued. We set out to write a Declaration of some sort and ended up writing an "Invitation to the Radical Center" instead. We decided it was time to declare an end to the hostilities that had consumed the issue of livestock grazing. But we also came to the conclusion that peace wasn't enough. To make progress and move forward we needed to mobilize the radical center—represented by the members of the New Ranch movement, for example—and by doing so give purpose, voice, and energy to an effort that had been growing by fits and starts. And we wanted to invite as many fellow "radical centrists" as possible to sign on.

Here is a portion of what we wrote:

We believe that how we inhabit and use the West today will determine the West we pass on to our children tomorrow; that preserving the biological diversity of working landscapes requires active stewardship; and that under current conditions the stewards of those lands are compensated for only a fraction of the values their stewardship provides.

We know that poor management has damaged land in the past and in some areas continues to do so, but we also believe appropriate ranching practices can restore land to health. We believe that some lands should not be grazed by livestock; but also that much of the West can be grazed in an ecologically sound manner. We know that management practices have changed in recent years, ecological sciences have generated new and valuable tools for assessing and improving land, and new models of sustainable use of land have proved their worth.

We therefore reject the acrimony of past decades that has dominated debate over livestock grazing on public lands, for it has yielded little but hard feelings among people who are united by their common love of land and who should be natural allies. We pledge our efforts to form the "Radical Center" where

- the ranching community accepts and aspires to a progressively higher standard of environmental performance;
- the environmental community resolves to work constructively with the people who occupy and use the lands it would protect;
- the personnel of federal and state land management agencies focus not on the defense of procedure but on the production of tangible results;
- the research community strives to make their work more relevant to broader constituencies;
- the land grant colleges return to their original charters, conducting and disseminating information in ways that benefit local landscapes and the communities that depend on them;
- the consumer buys food that strengthens the bond between their own health and the health of the land;
- the public recognizes and rewards those who maintain and improve the health of all land; and
- all participants learn better how to share both authority and responsibility.

Although we were pleased with our effort we wondered: could the radical center work in the real world? Would it make a difference where it mattered—on the ground and in people's lives?

Sterling and I bounced across Eagle Creek at a rocky ford and pulled up to a lonely white schoolhouse alongside a bevy of pickup trucks. Worn out, we opened the car doors and let the fresh breeze wash our faces. When Sterling asked why we were here, I explained that I had come to hear about a new group, the Upper Eagle Creek Watershed Association, and to make a short presentation myself. What I wanted to say was this: we were here to see the radical center in action.

The first thing I noticed inside the schoolhouse was a map. Like many streams in the American West, Eagle Creek is bounded by federal property—in this case by the U.S. Forest Service to the east and north, the BLM to the south, and the largely off-limits San Carlos Apache Indian Reservation to the west. The families of Eagle Creek felt hemmed in themselves as well. Challenges that I heard that morning included the declining economics of ranching, the increasing demands of environmentalists and recreationalists on public lands, the presence of Mexican wolves in their backyard, the possibility of the critically endangered southwestern willow flycatcher being discovered on their private land, the presence of an endangered fish in the creek, and a growing frustration with the slow, grinding bureaucratic machinery of the Forest Service and the U.S. Fish and Wildlife Service.

There was more.

"We didn't know our neighbors anymore," said Darcy Ely during her presentation. A third-generation rancher with a jovial smile, she was also the secretary of the Upper Eagle Creek Watershed Association (UECWA). "There were some new people we didn't know, and even the old ones were kinda' bunkered down. Some of us thought this wasn't healthy."

Fences in the valley were down or in need of repair. Homes were dark. The schoolhouse, which had fifteen students as recently as 1990, had shut down. Livestock had been removed from all of the ranches with public land allotments but two. Paychecks were scarce. The future looked bleak.

Then in 2003, Frank Hayes encouraged the families to attend a conference in Tucson, Arizona, organized by The Quivira Coalition featuring speakers from the Malpai group. A handful of ranch families went, discovering, ironically, that it look a long drive to a big city to get them to sit down together in one room.

"As I looked around the room," said UECWA president Chase Caldwell, an affable businessman from Chandler, Arizona, and landowner, "I recognized, maybe for the first time, that we had a huge resource of talent and experience gathered to work on our problem. Literally hundreds of years of experience in all types of business were there. This recognition of talent was an important revelation for all of us. These were friends and neighbors that I had known, but not really 'known.' It gave me a huge lift."

Back home on Eagle Creek, banking on this experience, and knowing that nonprofit organizations have access to government and foundation money that would otherwise be unavailable, they filed their hopes with the Internal Revenue Service. As part of the application for nonprofit status, they decided on a simple mission statement: UECWA is "an organization that benefits the people and the land."

They crafted four purposes for the organization:

- To work together as a community to preserve the heritage and traditions of Upper Eagle Creek
- To work together to improve and preserve the watershed and other valuable resources
- To work together to protect, enhance, and increase habitat for wildlife as well as domestic animals, especially in times of drought
- To work together to find a sustainable method of economic survival for the community

The group's PowerPoint presentation explained in detail how they planned to accomplish their goals—by partnering with a variety of wildlife groups, conservation organizations, and state and federal agencies to get the watershed back into shape, principally through the application of prescribed fire, which knocks back overgrown forests and grasslands. In fact, according to Kent Ellet, the district's range conservationist, one of the largest prescribed fires in the state of Arizona ever was being planned for a big portion of the Eagle Creek watershed with the cooperation of the ranchers. Without the watershed association this work may not have been possible.

"It was an out-of-the-box approach to a variety of concerns on the district," Frank Hayes told me in reference to their decision to form a nonprofit watershed group. "It was a response to a challenge I made to them to come up with a solution that avoided confrontation. I told them the Forest Service wanted to be a partner in the community, and now we are."

"Alone, it is doubtful that any one ranching operation can survive economically or socially by itself anymore," he continued. "Together, they can make a difference among themselves and as a group."

I gave a short presentation on the work of The Quivira Coalition, during which Sterling sat on my foot, much to the amusement of the audience. At the end of the day, we accepted the invitation of Twig and Shirley Winkle to spend the evening at their nearby ranch. An attractive young couple, new to ranching, and founding board members of UECWA, their experience captures both the challenges and the opportunities found in the watershed.

In the mid-1990s Twig and Shirley gave up city life in Mesa, Arizona, and purchased the historic Tule (pronounced Two-lee) Springs Ranch—fourteen thousand acres of rough, dry country, most of it public land, located at the southern end of the upper watershed. Their purchase put an exclamation point on a property whose history encapsulated much of the changing nature of ranching in the West over the past thirty years. In the mid-1970s, the long-time owner of the Tule became embroiled in a struggle with the federal government over his cattle operation on the national forest. Concerned about the deleterious effects of poor livestock management on the health of the land, the Forest Service cut the ranch's permit from 300 head of cattle down to 190. It was an almost unprecedented action for the time, and the rancher resisted mightily, taking the fight all the way to Washington, D.C.

Where he lost. When the Forest Service cut the permitted numbers to ninety head, because of continued concerns over poor cattle management, the rancher called it quits. By the time Twig and Shirley bought the deeded property, including the historic homestead, the number of cattle allowed to run on the forest allotment had been reduced to *fourteen.*

"It wasn't economical, to say the least," said Twig, with his easy laugh. "And the place was a mess too. But we were optimistic, or foolish, enough to give it a go."

Fortunately, Twig likes to fix things. His twenty years of experience as a heavy machinery mechanic in Mesa (where he met Shirley in Sunday school during the eighth grade), gave him both the muscles and a set of problem-solving skills that he would need to make the Tule work.

In the meantime they needed an income. They planted two gardens and began selling vegetables to folks in Safford. Shirley began raising purebred Airedale terriers for sale via the Internet. "Shirley makes more off her dogs than I do off my cows," said Twig, smiling again. They also remodeled one of the old buildings into a guest house for outfitters, vacationers, and other folks interested in a remote ranch experience.

But the process of trying to open a dude ranch for business had been sobering.

"We got clobbered by the insurance," said Shirley, "especially on the horses. It came out of the blue and kinda' depressed us."

In the meantime, Twig began to earn a paycheck for an unusual type of activity for a rancher: monitoring the land. Knowing the Forest Service was shorthanded, and that his neighbors would need data in support of continued livestock production on public land, Twig went "back to school" and became proficient at grass identification, transects, photography, and reading the signs of erosion.

"The great thing about monitoring is that you learn about limitations," said Twig. "You get a real good sense of exactly how far you can go, and no further."

His education took them all over the Southwest, from riparian monitoring training near Silver City, New Mexico, to a low-stress livestock handling clinic at Ghost Ranch (the same one where I met Steve Allen and Guy Glosson), near Santa Fe. They absorbed new knowledge like sponges. In fact, when they began to apply the principles of progressive ranch management to the Tule, in combination with the data from their own monitoring, the Forest Service responded with a proposal to raise their permitted number of cattle from fourteen to fifty-five head—a proposal that pleased the Winkles even if it didn't solve their financial situation.

"Even with the increase, the ranch is still not economical from a cattle standpoint," said Twig, "But that's okay because what's important is I'm getting paid for good stewardship." To that end, Twig and Shirley are taking the lead on the idea of a communal herd of cattle in the Eagle Creek watershed. By combing three herds into one they can reduce the costs of the ranch operation by consolidating expenses and labor while improving the health of the grasses, by giving them plenty of time to rest between grazing episodes.

"Ranchers always say that if they take care of the land, it will take care of them," said Twig. "And they're right. But today you've got to monitor to demonstrate it. And the cattle side of things has also got to be profitable. I know we can do both."

I learned from the Winkles that UECWA's gamble had begun to pay off. The watershed group had already received three grants: one for $40,000 from the Arizona Heritage Fund, through the Arizona Game and Fish Department; one from a group of wildlife associations for $50,000; and one through members Jan and Will Holder from the nonprofit Sonoran Institute for $7,500.

The first two grants were developed in cooperation with the Forest Service to maintain and develop trail systems in the forest and to continue an ongoing ecosystem restoration project through prescribed burning and mechanical thinning of trees. The third one, from the Sonoran Institute, was a planning grant. The goal was to reform the economic base of the Eagle Creek community through land improvement projects, such as riparian protection, monitoring, forest thinning and burning, and water development.

Plans are one thing, of course, implementing them another. The challenges are varied and daunting, including the occasional "people problems" typical of any organization. "It's easier to fix the land sometimes than it is to hold the human equation together," said Twig. "But we've done alright so far."

Then, in the summer of 2005, came the first in a series of successful grant applications after the 2004 schoolhouse meeting, all building on the group's watershed restoration plan they composed with the Forest Service:

- $900,000 from the Arizona Department of Agriculture to be divided among the families for ranch and habitat improvement
- $360,000 from the Arizona Department of Environmental Quality (via the U.S. Environmental Protection Agency's 319 Clean Water program) for watershed restoration work
- $10,000 from a community-assistance grant in 2006 to buy satellite phones for better communication within and outside of the watershed
- $200,000 in matching grants made to the Clifton District by the regional office of the Forest Service for restoration activities
- $850,000 in 2007 in a second round of grants from the Arizona

Department of Agriculture for continued improvements in the watershed

Successes kept coming. In 2006, the national Society for Range Management gave Kent Ellet, the district range conservationist, one of their annual awards; and Kent and Frank received a nationwide award from the Forest Service.

The ranchers received a different sort of recognition. According to Chase Caldwell, in 2004 the number of cattle that were allowed to run on the national forest allotments was 200, out of a permitted potential of 1,800 for the whole watershed. By 2007, the actual number of head on the ground had been raised to nearly 700. "The monitoring supported the increase," said Chase, who is about to return his own cattle to national forest land. "As did all the improvements we've accomplished on the land."

Speaking of on-the-ground, here is what has been accomplished or is pending as of fall 2007:

- 17,000 acres of prescribed fire completed
- 1,500 acres of hand-thinned forest restoration completed
- 30 miles of fences reconstructed
- 25 miles of waterlines repaired or installed
- Over 50,000 acres of prescribed fire set to go
- 4,000 more acres of hand-thinned forest work set to go
- Tons of goodwill created among landowners, wildlife agencies, conservation groups, the Forest Service, county commissioners, and neighbors

The most recent sign of success, said Chase, was the desire of thirty or so ranchers in the Blue River watershed, to the east, to join UECWA, potentially expanding the group's geographic impact significantly. "We think it's a great opportunity," said Chase, "though it will probably mean we'll need to change our name."

The radical center, in other words, seemed to be holding.

"We can see momentum building and we have hope for our future as a community and the future of ranching," he continued. "It's a testament to what a determined group of people can accomplish."

This recalled the famous quote by Margaret Mead, the American anthropologist, who said: "Never doubt that a small group of thoughtful, committed citizens can change the world. Indeed, it is the only thing that ever has."

Actually, it recalled two more quotes as well. The first came to me on the drive home from Eagle Creek in 2004. Pulling onto the winding highway, I slipped a favorite CD of Sterling and Olivia's into the car's stereo. As we approached the big mine a while later, I caught a snippet of a song titled *What's Mine Is Yours* sung by a bear who lived in a big blue house. The words went like this: *"We're in the same boat, let's share the oars. You know, we'll go much faster if we do."*

I glanced over at Sterling, who was looking out the window. I wanted to tell him: it's *your* boat too. Someday you'll be pulling the oars. Instead, I recalled an observation by Wallace Stegner that I had read many years ago. Commenting on the prospects of his cherished American West in his collection of essays *The Sound of Mountain Water,* he wrote: "This is the native home of hope. Only when it fully learns that cooperation, not rugged individualism, is the pattern that most characterizes and preserves it, then it will have achieved itself and outlived its origins. Then it has a chance to create a society to match the scenery."

Reaching a straight stretch of open road, I reached out and stroked my son's hair. For Sterling's sake, as well as that of his generation, I hope we're on the way.

PART TWO

The Working Wilderness

Chapter Six

THE WORKING
WILDERNESS

> The only progress that counts is that on the actual
> landscape of the back forty.
>
> ALDO LEOPOLD

U Bar Ranch *Silver City, New Mexico*

During a conservation tour of the well-managed U Bar Ranch near
Silver City, New Mexico, I was asked to say a few words about a map a
friend had recently given to me.

We were taking a break in the shade of a large piñon tree, and I rose
a bit reluctantly (the day being hot and the shade being deep) to explain
that the map was commissioned by an alliance of ranchers concerned
about the creep of urban sprawl into the five-hundred-thousand-acre
Altar Valley, located southwest of Tucson, Arizona. What was different
about this map, I told them, was what it measured: indicators of rangeland health. Our tour, in fact, had been designed to study signs of range
health, such as grass cover (positive) and bare soil (negative), and what
they might tell us about livestock management in arid environments.
What they told us was this: the U Bar was in pretty good condition.

What was important about the map, I continued, was what it said about a large watershed. Drawn up in multiple colors, the map expressed the intersection of three variables: soil stability, biotic integrity, and hydrological function—soil, grass, and water, in other words. The map displayed three conditions for each variable—"stable," "at risk," and "unstable"—with a color representing a particular intersection of conditions. Deep red designated an unstable, or unhealthy, condition for soil, grass (vegetation), and water, for example, while deep green represented stability in all three. Other colors represented conditions between these extremes.

In the middle of the map was a privately owned ranch called the Palo Alto. Visiting it recently, I told them, I was shocked by its condition. It had been overgrazed by cattle to the point of being nearly "cowburnt," to use author Ed Abbey's famous phrase. As one might expect, the Palo Alto's color on the map was blood red and there was plenty of it.

I paused briefly; now came the controversial part. This big splotch of blood red continued well below the southern boundary of the Palo Alto, I said. However, this was not a ranch, but part of the Buenos Aires National Wildlife Refuge, a large chunk of protected land that had been cattle-free for nearly sixteen years. . . .

That was as far as I got. Taking offense at the suggestion that the refuge might be ecologically unfit, a young woman from Tucson cut me off. She knew the refuge, she explained, having worked hard as a volunteer with an environmental organization to help "heal" it from decades of abuse by cows.

The map did not blame anyone for current conditions, I responded; nor did it offer opinions on any particular remedy. All it did was ask a simple question: Is the land functioning properly at the fundamental level of soil, grass, and water? For a portion of the Buenos Aires National Wildlife Refuge the answer was "no." For portions of the adjacent privately owned ranches, which were deep green on the map, the answer was "yes."

Why was that a problem?

I knew why. I strayed too closely to a core belief of my fellow conservationists: that "protected" areas, such as national parks, wilderness areas, and wildlife refuges, must always be rated, by definition, as being in better ecological condition than adjacent "working" landscapes.

Yet the Altar Valley map challenged this paradigm at a basic level, and when the tour commenced again on a ranch that would undoubtedly encompass more deep greens than deep reds on a similar map, I saw in the reaction of the young activist a reason to rethink the conservation movement in the American West.

From the ground up.

CS Ranch *Cimarron, New Mexico*

My observation received a boost a few weeks later while sitting around a campfire after a tour of the beautiful, one-hundred-thousand-acre CS Ranch, located in northeastern New Mexico. Staring into the flames, I found myself thinking about ethics. I believed at the time, as do many conservationists, that the chore of ending overgrazing by cattle in the West was a matter of getting ranchers to adopt an ecological ethic along the lines Aldo Leopold suggested in his famous "Land Ethic" essay, in which he argued that humans had a moral obligation to be good stewards of nature.

The question, it seemed to me, was how to accomplish this lofty goal.

I decided to ask Julia Davis-Stafford, our host, for advice. Years earlier, Julia and her sister Kim talked their family into switching to planned grazing, a decision that over time caused the ranch to flourish economically and ecologically. In fact, the idea for my query came earlier that day when I couldn't decide which was more impressive: the sight of a new beaver dam on the ranch or Julia's strong support for its presence.

The Davis family, it seemed to me, had embraced Leopold's land ethic big time. So, over the crackle of the campfire, I asked Julia, "How do we get other ranchers to change their ethics, too?"

Her answer altered everything I had been thinking up until that moment.

"We didn't change our ethics," she replied. "We're the same people we were fifteen years ago. What changed was our knowledge. We went back to school, in a sense, and we came back to the ranch with new ideas."

Knowledge *and* ethics, neither without the other, I suddenly saw, are the key to good land stewardship. Her point confirmed what I had observed during visits to livestock operations across the region: many ranchers *do* have an environmental ethic, as they have claimed for so long. Often their ethic is a powerful one. But it has to be matched with *new* knowledge—especially ecological knowledge—so that an operation can adjust to meet changing conditions, both on the ground and in the arena of public opinion. Of course, a willingness on the part of a rancher to "go back to school" is a prerequisite to gaining new insights. Tradition, however, seemed to have a lock on many ranchers.

The same thing is true of many conservationists. Tradition was just as much an obstacle in the environmental community as it was in agriculture. It wasn't just the persistence of various degrees of bovine bigotry among activists, despite examples of healthy grazed landscapes like the U Bar, either. It was more the stubborn belief in a hands-off relationship between humans and nature expressed in the long-standing dualism of environmentalism that said recreation and play in nature were acceptable while work and use were not.

If conservationists went back to school, as the Davis family did, what could we learn? Aldo Leopold had a suggestion that can help us today: study the fundamental principle of "land health," which he described as "the capacity of the land for self-renewal," with conservation being "our effort to understand and preserve this capacity."

By studying the elements of land health, especially as they change over time, conservationists could learn that grazing is a natural process. The consumption of grass by herbivores in North America has been going on for millions of years—not by cattle, of course, but by bison, elk, deer, (and grasshoppers, rabbits, even ants)—resulting in a complex relationship between grass and grazer that is ecologically self-renewing. We could learn that a re-creation of this relationship with domesticated cattle lies at the heart of the new ranching movement, which is why many progressive ranchers think of themselves as "grass farmers" instead of beef producers.

We could also learn that many landscapes need periodic pulses of energy, in the form of natural disturbance, such as fires and floods (but not the catastrophic kind), to keep things ecologically vibrant. Many conservationists know that low-intensity fires are a beneficial form of disturbance in ecosystems because they reduce tree density, burn up old grass, and aid nutrient cycling in the soil. But many of us don't know that small flood events can be a positive agent of change too, as can drought, wind storms, and even insect infestation. We also may not know that animal impact caused by grazers, including cattle, can be a beneficial form of disturbance.

We could further learn, as the Davis family did, that the key to healthy "disturbance" with cattle is to control the timing, intensity, and frequency of their impact on the land. The CS, and other progressive ranches, bunch their cattle together and keep them on the move, rotating the animals frequently through numerous pastures. Ideally, under this system no single piece of ground is grazed by cattle more than once a year, thus ensuring plenty of time for the plants to recover. The keys are regulating where cattle go, which can be done with fencing or a herder, and the timing of their movement, in which the herd's moves are carefully planned and monitored. In fact, as many ranchers have learned, overgrazing is more a function of timing than it is of numbers of cattle. For example, imagine the impact 365 cows would have in one day of grazing in one pasture—now imagine what one cow would do in 365 days of grazing in the same pasture. Which is more likely to be overgrazed? (Hint: have you ever seen what a backyard lot looks like after a single horse has grazed it for a whole year?)

We could also learn, as I did, that much of the damage we see today on the land is historical—a legacy of the "Boom Years" of cattle grazing in the West. Between 1880 and 1920 millions of hungry animals roamed uncontrolled across the range, and the overgrazing they caused was so extensive, and so alarming, that by 1910 the U.S. government was already setting up programs to slow and to heal the damage. Today, cattle numbers are down, way down, from historic highs, a fact not commonly voiced in the heat of the cattle debate.

A willingness to adopt new knowledge allowed the Davis family to maintain their ethic yet stay in business. Not only did it improve their bottom line, it helped them meet evolving values in society, such as a rising concern among the pubic about overgrazing. Rather than fight change, they had switched.

As the embers of the campfire burned softly into the night, I wondered if the conservation movement could do the same.

Kaibab National Forest *Flagstaff, Arizona*

My friend Dan Dagget likes to tell a story about a professor of environmental studies he knows who took a group of students for a walk in the woods near Flagstaff, Arizona. Stopping in a meadow, the professor pointed at the ground and asked, not so rhetorically, "Can anyone tell me if this land is healthy or not?" After a few moments of awkward silence, one student finally spoke up, "Tell us first if it's grazed by cows or not?" In a similar vein, a Santa Fe lawyer told me that a monitoring workshop at the boundary between a working ranch and a wildlife refuge south of Albuquerque had completely rearranged his thinking. "I've done a lot of hiking and thought I knew what land health was," he said, "but when we did those transects on the ground on both sides of the fence, I saw that my ideas were all wrong."

These two instances illustrate a recurring theme in my experience as a conservationist. To paraphrase a famous quote by a Supreme Court justice, members of environmental organizations "can't define what healthy land is but they know it when they see it."

The principal problem is that we are "land illiterate." When it comes to "reading" a landscape, we might as well be studying a foreign language. Many of us who spend time on the land don't know our perennials from our annuals, what the signs of poor water cycling are, what leads to deeply eroded gullies, or, simply by looking, whether a meadow is healthy or not.

For a long time this situation wasn't our fault. What all of us lacked—

rancher, conservationist, range professional, curious onlooker—was a common language to describe the common ground below our feet. But that has changed.

In recent years, range ecologists have reached a consensus on a definition of health: the degree to which the integrity of the soil and ecological processes of rangeland ecosystems are sustained over time. Components include water and nutrient cycling, energy flow, and the structure and dynamics of plant and animal communities. In other words, when scarce resources such as water and nutrients are captured and stored locally, by healthy grass plants, for example, then ecological integrity can be maintained and sustained. Without these resources—if water runs offsite instead of percolating into the soil, or grass plants die due to excessive erosion of the topsoil, for example—this integrity will likely be lost over time, perhaps quickly.

This is the language of soil, grass, and water.

Taking it the next step, range ecologists echo Aldo Leopold's famous quote, "Healthy land is the only permanently profitable land." Producing commodities and satisfying values from a stretch of land on a sustained basis, they insist, depends on the renewability of internal ecological processes. In other words, before land can sustainably support a value, such as livestock grazing, hunting, recreation, or wildlife protection, it must be functioning well at a basic ecological level. Before we, as a society, can talk about designating critical habitat for endangered species, or increasing forage for cows, or expanding recreational use, we need to know the answer to a simple question: is the land healthy at the level of soil, grass, and water?

If the answer is "no," then all our values for that land may be at risk.

Or as Kirk Gadzia, coauthor of *Rangeland Health*, the pioneering work published in 1994 by the National Academy of Sciences, likes to put it, "It all comes down to soil. If it's stable, there's hope for the future. But if it's moving, then all bets are off for the ecosystem." It is a sentiment Roger Bowe, an award-winning rancher from eastern New Mexico, echoes. "Bare soil is the rancher's number one enemy."

It should become the number one enemy of conservationists as well.

The publication of *Rangeland Health* was the touchstone for a new consensus on the meaning of land health within the scientific and range professional communities. It paved the way for the debut, in 2000, of a federal publication entitled *Interpreting Indicators of Rangeland Health,* which provides a seventeen-point checklist for the qualitative assessment of upland health. A similar assessment has been made of stream health by a federal interagency group known as the National Riparian Team. The indicators of land health include measures of the presence of rills (small gullies), bare ground (lack of vegetation), pedestaling (grass plants left high and dry by water erosion), litter (dead grass, which retards the erosive impact of rain and water), soil compaction (which can prohibit water infiltration), plant diversity (generally a good thing), and invasive species (generally not)—the same indicators that formed the basis of the Altar Valley map that I described on the tour.

Scientists at the USDA's Jornada Experimental Range, near Las Cruces, New Mexico, published a peer-reviewed protocol for quantitatively measuring rangeland health—the next step after an assessment. Using a methodology that quantifies a watershed's ability to resist degradation, as well as recover from disturbance, this protocol, according to the manual, "is designed to quantify the potential of the system to function to support a range of societal values rather than to support any particular value." Healthy land, in other words, supports many values while unhealthy land offers diminishing support for values over the long run.

This was the message I tried to communicate to the young activist under the tree that hot summer day—that a rangeland health paradigm, employing standard indicators, allows all land to be evaluated equally and fairly. By adopting it, the conservation movement could begin to heed Aldo Leopold's advice that any activity that degrades an area's "land mechanism," as he called it, should be curtailed or changed, while any activity that maintains, restores, or expands it should be supported. It should not matter if that activity is ranching or recreation.

Chaco Culture National Historical Park

Southeast of Farmington, New Mexico

In an attempt to understand the issues of land health better, I paid a visit to a famous fenceline contrast. This particular fence separated the Navajo Nation, and its cows, from Chaco Culture National Historical Park, a World Heritage site and archaeological preserve located in the high desert of northwest New Mexico. Cattle-free for over fifty years, Chaco's ecological condition became a pedagogical topic some years ago when biologist Allan Savory used the boundary to highlight the dangers of too much rest from the effects of natural disturbance, including grazing and fire.

I wasn't a big fan of fenceline contrasts, mostly because I dislike dichotomies: grazed/ungrazed, wild/unwild, either/or, us/them. The world is more complicated than that. I'd rather take fences down, or move beyond them. But fenceline contrasts have educational value, especially for new students of range health—like me. I wanted to see this contrast in particular, but I knew I needed help interpreting what I saw, so I asked Kirk Gadzia to come along.

Both of us were well aware of the park's history—that a century of overgrazing by livestock had badly degraded the land surrounding the famous ruins. We also understood that the era's typical response to this legacy of overuse was to try and protect the land from further degradation, principally with the tools of federal ownership and a barbed wire fence. That's how Chaco became a national park. At the time, it was a common and probably appropriate scenario played out all across the West. But Kirk and I didn't go to Chaco to argue with history, or pick a fight with the National Park Service. We weren't there to offer "solutions" to any particular problem either. We simply wanted to take the pulse of the land on both sides of a fence.

We stopped along the road at the eastern boundary of the park (this was during the summer growing season). On the Chaco side we saw a great deal of bare ground, as well as many forbs, shrubs, and other woody material, some of it dead. We saw few young plants, few perennial or bunchgrasses, lots of wide spaces between plants, lots of oxidized plant matter (dead grass turning gray in the sunlight), and a great deal of poor plant vigor. We saw both undisturbed, capped soil (without a cover of grass or litter, soil will often "cap" or seal when exposed to pounding rain, thus preventing seed germination) and abundant evidence of soil movement, including gullies and other signs of erosion. Among the positives, we saw a greater diversity of plant species than on the Navajo side, more birds, more seed production, and no sign of overgrazing.

On the Navajo side we saw lots of plant cover and litter, lots of perennial grasses, tight spaces between plants, few woody species, a wide age-class distribution among the plants, little evidence of oxidization, and lots of bunchgrasses. We saw little evidence of soil movement, no gullies, and far fewer signs of erosion than on the Chaco side. On the other hand, we saw less species diversity, a great deal of compacted soil, fewer birds, less seed production, a great deal of manure, and numerous signs of overgrazing.

"So, which side is healthier?" I asked Kirk.

"Neither one is healthy, really," he replied, "not from a watershed perspective anyway." He noted that the impact of livestock grazing on the Navajo side was heavy; plants were not being given enough time to recover before being bitten again (Kirk's definition of overgrazing). As a result, the plants lacked the vigor they would have exhibited in the presence of well-managed grazing.

However, Kirk thought the Chaco side was in greater danger, primarily because it exhibited major soil instability due to gullying, capped soil, and lack of plant litter. "The major contributing factor to this condition is the lack of tightly spaced perennial plants," he continued, "which exposes the soil to the erosive effects of wind and rain. When soil loss is increased, options for the future are reduced."

"But isn't Chaco supposed to be healthier because it's protected from grazing?" I asked.

"That's what people always seem to assume," said Kirk. "In my experience in arid environments around the world, total rest from grazing has predictable results. In the first few years, there is an intense response in the system as the pressure of overgrazing is lifted. Plant vigor, diversity, and abundance often return at once and all appears to be functioning normally. Over the years, however, if the system does not receive periodic natural disturbance, by fire or grazing, for example, then the overall health of the land deteriorates. And that's what we are seeing on the Chaco side."

Then he added a caveat.

"Maybe land health isn't the issue here," he said. "It may be more about values. Is rest producing what the park wants? Ecologically, the answer is probably 'no.' But from a cultural perspective, the answer might be 'yes.' From the public perspective too. People may not want to see fire or grazing in their park."

But at what price, I wondered? Later in the day we learned that the National Park Service was so worried about the threat of erosion to Chaco's world-class ruins that they intended to spend a million dollars constructing an erosion control structure in Chaco Wash. This told us the agency knows it has a "functionality" crisis on its hands.

But how can proper functioning condition be restored if the Park Service's hands are tied by a cultural value that says Chaco must be "protected" from incompatible activities, even those that might have a beneficial role to play in restoring the park lands to health?

As I drove home, I realized that this tension between value and function at Chaco was a sign of a new conflict spreading slowly across the West, symbolized by a fence. The cherished protection paradigm, embedded in the conservation movement since the days of John Muir, rubbed against something new, something energetic, something beyond the fence.

Bandelier National Monument *Near Los Alamos, New Mexico*

The passage of the Wilderness Act in 1964 was a seminal event in the history of the American conservation movement. For the first time, wilderness had a legal status, enabling the designation and the protection of "wildland" that had been under siege in that era of environmental exploitation. Energized, the conservation movement grabbed the wilderness bull by both horns and has not let go to this day. But the act's passage also had an unforeseen consequence—it set in motion the modern struggle between value and function in our western landscapes.

This tension took a while to develop. In 1964, there was intellectual harmony between the social and ecological arguments for the creation of a federal wilderness system. No reconciliation was necessary between the act's definition of wilderness as a tract of land "untrammeled by man . . . in which man is a visitor who does not remain" and Aldo Leopold's declaration, published in *A Sand County Almanac* fifteen years earlier, that wilderness areas needed protection because they were ecological "base datums of normality."

Leopold asserted that wilderness was "important as a laboratory for the study of land health," insisting that in many cases "we literally do not know how good a performance to expect of healthy land unless we have a wild area for comparison with sick ones." Author Wallace Stegner extended the medical metaphor when he argued that wilderness was "good for our spiritual health even if we never once in ten years set foot in it."

But a lot has changed in the years since the passage of the Wilderness Act. Although most Americans still believe wilderness is necessary for social and mental health, few ecologists now argue that wilderness areas can be considered as "base datums" of ecological health.

For example, an article titled "Would Ecological Landscape Restoration Make the Bandelier Wilderness More or Less of a Wilderness?" was published in the journal *Wild Earth* in 2001. The authors, including ecologist Craig Allen, who has studied Bandelier National Monument, located in north-central New Mexico, for nearly twenty years, stated matter-of-factly that "most wilderness areas in the continental United States are not pristine, and ecosystem research has shown that conditions in many are deteriorating."

In the authors' opinion, the Bandelier Wilderness is suffering from "unnatural change" as a result of historic overuse of the area in the late nineteenth and early twentieth centuries—grazing by sheep principally—which triggered unprecedented change in the park's ecosystems, resulting in degraded and unsustainable conditions. "Similar changes," they wrote, "have occurred throughout much of the Southwest."

Specifically, soils in Bandelier are "eroding at net rates of about one-half inch per decade. Given soil depths averaging only one to two feet in many areas, there will be loss of entire soil bodies across extensive areas." This is bad because the loss of topsoil, and the resulting loss of water available for plants, impedes the growth of all-important grass cover, thus reducing the incidence of natural and ecologically necessary fires.

The reduction and eventual elimination of livestock grazing, starting with the creation of the park in the 1930s, was no panacea for Bandelier's functionality crisis, however. Herbivore exclosures established in 1975 show that protection from grazing, by itself, "fails to promote vegetative recovery." Without management intervention, the authors argued, this human-caused case of accelerated soil erosion will become irreversible. "To a significant degree the park's biological productivity and cultural resources are literally washing away."

Their summation is provocative: "We have a choice when we know land is 'sick.' We can 'make believe,' to quote Aldo Leopold, that everything will turn out alright if Nature is left to take its course in our unhealthy wildernesses, or we can intervene—adaptively and with humility—to facilitate the healing process."

I believe new knowledge about the condition of the land leaves us no choice: we must intervene. However, this turns a great deal of "old" conservation thinking on its head.

For instance, Wallace Stegner once wrote, "Wildlife sanctuaries, national seashores and lakeshores, wild and scenic rivers, wilderness areas created under the 1964 Wilderness Act, all represent a strengthening of the decision to hold onto land and manage large sections of the public domain rather than dispose of them *or let them deteriorate.*" [emphasis added]

But we have let them deteriorate, as the Buenos Aires, Chaco, and Bandelier examples demonstrate. Whether their deteriorated condition is a result of historical overuse or some more recent activity is not as important as another question: what are we going to do to heal land we know to be sick?

Clearly it's not 1964, or 1946, anymore. The harmony between value and function in the landscape, including in some of our "protected" places, has deteriorated along with the topsoil. This functionality crisis raises important questions for all of us. What, for instance, are the long-term prospects for wildlife populations in the West, including keystone predator species, if the ecological integrity of these special places is being compromised at the level of soil, grass, and water? Also, does "protection" from human activity preclude intervention, and if so, at what cost to ecosystem health? And on a larger scale, how do we "protect" our parks and wildernesses from the effects of global warming, acid rain, and noxious weed invasion?

Furthermore, the dualism of "protected" versus "unprotected" creates a stratification of land quality and land use that bears little relation to land health. As conservationist Charles Little wrote, "Leopold insisted on dealing with land whole: the system of soils, waters, animals, and plants that make up a community called 'the land.' But we insist on discriminating. We apply our money and our energy in behalf of protection on a selective basis." He went on to say, "The idea of a hierarchy in land quality is *the* tenet of the conservation and environmental movement."

Since John Muir's day, the conservation movement has based this hierarchy on the concept of "pristineness"—the degree to which an area of land remains untrammeled by humans. As late as 1964, when not as much was known about ecology or the history of land use, it was still possible to believe in the pristine quality of wilderness as an ecological fact, as Leopold did. Today, however, pristineness must be acknowledged to be a value, something that exists mostly in the eye of the beholder.

Biologist Peter Raven put it in blunt ecological terms: "There is not a square centimeter anywhere on earth, whether it is in the middle of the Amazon basin or the center of the Greenland ice cap, that does not receive every minute some molecules of a substance made by human beings."

I believe the new criterion should be *land health*. By assessing land by one standard, a land health paradigm encourages an egalitarian approach to land quality, thereby reducing conflicts caused by clashing cultural values (theoretically, anyway). By employing land health as the common language to describe the common ground below our feet, we can start fruitful conversations about land use rather than resort to the usual dualisms that have dominated the conservation movement for decades. We can also gain new *knowledge* about the condition of a stretch of land, knowledge that can help us make informed decisions.

I know a chunk of Bureau of Land Management (BLM) land west of Taos, New Mexico, that will never be a wilderness area, national park, or wildlife refuge. It is modest land, mostly flat, covered with sage, and very dry. In its modesty, however, it is typical of millions of acres of public land across the West. It is typical in another way too: it exists in a degraded ecological condition, the result of historic overgrazing and modern neglect. A recent qualitative land health assessment revealed its poor condition in stark terms (lots of bare soil, many signs of erosion, and a lack of plant diversity), confronting us with the knowledge that more than forty years of total rest from livestock grazing had not healed the land. Some of it, in fact, teetered on an ecological threshold, threatening to transition to a more deeply degraded state.

Fortunately, as humble and unhealthy as this land is, it is not unloved. The wildlife like it, of course, but so do the owners of the private land intermingled with the BLM land, some of whom built homes there. The area's two new ranchers also have great affection for this unassuming land and want to see it healed.

These ranchers are using cattle as agents of ecological restoration. Through the effect of carefully controlled herding, they intend to trample the sage and bare soil, much of which is capped solid, so that native grasses can reestablish again. The ranchers are calling this act of restoration a "poop-and-stomp," and its effects are being carefully monitored using the new land health protocols.

Using cattle as agents of ecological restoration is not as novel as it may sound. In fact, in his 1933 classic book *Game Management*, Aldo Leopold wrote more generally that wildlife "can be restored with the same tools that have hithertofore destroyed it: fire, ax, cow, gun, and plow."

The difference, of course, is the management of the tool, as well as the goals of the tool user.

I believe conservationists should share the same goal as these ranchers: to transform red to green on maps such as that of the Altar Valley and the land west of Taos. Whether we use cattle or some other method of restoration, the result must be a thousand acts of healing, starting at the level of soil, grass, and water. And healing must extend to communities of people as well, both urban and rural. Restoration jobs could be a boon to local economies; and volunteers from environmental groups could help. Turning red to green could unite us, no matter what our values.

By developing a common language to describe the common ground below our feet, by working collaboratively to heal land and restore rural economies, by monitoring our progress scientifically, and by linking function to value in a constructive manner, a land health paradigm can steer us toward fulfilling Wallace Stegner's famous dream of creating a "society to match the scenery."

West Elks Wilderness *Near Paonia, Colorado*

Wanting to learn more about the compatibility between well-managed ranching and wilderness values, I had the privilege one summer of riding a horse into the West Elks Wilderness, high in the mountains above Paonia, Colorado, with rancher Steve Allen and Forest Service range conservationist Dave Bradford, whom I met at the herding clinic I organized in 1999.

After buying a ranch in the late 1980s, Steve, like the Davis family, went "back to school" to learn the principles of progressive cattle management, including the details of low-stress livestock herding. Upon "graduation" he convinced five other neighboring ranchers with grazing permits in the West Elks to form a pool and begin herding their cattle as one unit through the mountains. They convinced the Forest Service to let them give it a try.

Pool riders guide the thousand-head herd of cattle through a long arc in the mountains with the aid of border collies and the occasional temporary electric fence. They move the herd every ten days or so, which allows the land plenty of time to recover. Because traditional fences are no longer necessary, the ranchers voluntarily removed hundreds of miles of barbed wire fence in the wilderness, a boon to wildlife and backpackers alike.

Dave Bradford went to the same school as Steve and came back determined to quantify the effects of this new thinking. As I learned, he rides the range frequently, reads monitoring transects on a regular basis, and publishes the results. He also has done quite a bit of historical research, including uncovering "before" photographs of the range, to gain new knowledge on the conditions of the land.

Also joining us that morning was Tara Thomas, the new executive director of a Paonia-based conservation organization, whom I had invited along. The support of her predecessor for the West Elks herding experiment had been crucial to its early success, and I was curious what she thought as an heir to the project. She was curious too; she explained that she had recently backpacked the very trail we were riding that morning.

What we saw shocked us at first. The herd of cattle had moved along the trail just days before, beating it into a muddy pulp. It looked like a tornado had touched down; shattered brush and trampled grass were ubiquitous, as was the cow poop. It certainly was not your standard Sierra Club–calendar image of wilderness.

"This looks great!" yelled Dave, as we climbed a steep hill on horseback. "Look at all this disturbance. Come back here in a month and you would never know the cattle went through here, it'll be so lush."

I turned to Tara. "People call me all the time and complain," she said. "They're hikers. They don't think there should be cows in the wilderness."

"What do you tell them?" I asked.

"I tell them it's a working wilderness," she replied, spurring her horse forward.

Steve led us to a high meadow where we found a small herd of cattle that had broken off from the main pack. After lunch we would spend the rest of the day driving the cattle back down the mountain in a chaotic rush of snapping branches, surging adrenaline, and hard work.

Before we began, however, we all sat in the green meadow and ate lunch among the blooming wildflowers, admiring the view. Each of us, rancher, federal manager, and activist, shared the same thought: what a treasure this land is! Sitting there, I was reminded of why I became a conservationist—to explore the solace of open spaces; to look and learn, and teach in turn; to celebrate cultural diversity alongside biological diversity; and to revel in nature's model of good health.

And to try to understand, as John Muir did, that every part of the universe is hitched to everything else.

Chapter Seven

GOALS THAT UNITE

Santa Barbara, California

When Dan Dagget gave a talk at the annual Bioneers Conference, near San Francisco, which is a three-day celebration of environmental activism, antiestablishment politics, and bio-friendly approaches to food and land management, he began by asking audience members if they had taken care of their environmental responsibilities that day. Had any of them gone hunting in a pack? Started a grass fire? Piled rocks in a gully? Chased any bison off a cliff?

In response, some people jumped to their feet and walked out of the auditorium.

This didn't surprise the former Earth First! activist. Dan has been causing people discomfort since the early 1970s, when he fought the coal mining industry in his native southeastern Ohio. After moving to Flagstaff, Arizona, he became active in efforts to stop proposed uranium mines near the Grand Canyon. He also worked to overturn Arizona's "stockkiller" law, which essentially gave ranchers free rein to kill mountain lions and black bears. He even engaged in a stunt in which he confronted a state legislator with a huge bear trap that he carried to the legislator's office, springing it as TV cameras rolled. For these efforts, Dan was designated as one of the nation's top 100 activists by the Sierra Club in 1992.

Three years later, however, it was his fellow environmentalists who were discomfited with Dan's Pulitzer Prize–nominated book *Beyond the Rangeland Conflict: Toward a West that Works*. That's because it gave lie to the long-standing idea that environmentalists and ranchers in the American West held incompatible goals. In it he profiled a dozen or so "New Ranchers" around the West—though Dan didn't use that term— who were employing progressive land stewardship methods successfully. It earned him a great deal of enmity among his former colleagues. In fact, he was effectively drummed out of the movement.

After reading his book in 1996 I immediately called Dan with a question: what had caused him to change his mind? Why did he switch from street brawler to peace maker? His answer surprised me.

"I didn't change," he replied. "I'm still an environmentalist, or at least that's how I think of myself. My goals are the same ones they were when I was fighting those strip mines in Ohio. I'm the same person. Only my approach has changed."

We kept talking. He told me his approach first began to change back in 1989, when he accepted an invitation from a fellow activist to join a facilitated meeting between ranchers and environmentalists in a carport in Phoenix. Six of each showed up, suggesting the name "6-6" for the group, for "six of them and six of us," he said. The subject of the meeting was the state's "stockkiller" law, which Dan's friend saw as an obstacle to her dream of reintroducing the endangered Mexican wolf to Arizona. It was an explosive topic, and Dan went ready to rumble. "My intent was to bash the other side and do everything I could to win the inevitable confrontation," he told me.

But something funny happened once inside the gladiator's ring. The facilitator, Tommie Martin, who came from a ranching family, opened the meeting by asking everyone in the carport to describe their ideal rangeland. What did they want the land to look like? What did they want its condition to be? What did they want their relationship to this land to be like? People raised their hands. Dan said what happened next shocked him. "Ranchers started saying things that I had planned to say, things that I thought would only come from an environmentalist, a radical one at that: more open space, less people, more places in which a person could get lost, healthy populations of wildlife, restored streams with water running in them, and ways for people to live in the West that didn't destroy the very reason most of us had moved here in the first place."

They were descriptive ends, Dan noticed, instead of prescriptive means. Instead of saying "get rid of the cows" or "keep the wolves out"—prescriptions that would have instantly led to confrontation—the twelve participants, guided gently by Tommie, described their vision for the land itself, discovering they wanted many of the same things, much to their surprise.

Dan was so impressed by what he heard the ranchers say at the first 6-6 meeting, especially about how some of them managed their land, that he decided to hit the road, visiting many of the ranches that he would eventually profile in *Beyond the Rangeland Conflict*. He visited ranchers in Nevada, Idaho, and Mexico that were revitalizing grasslands with their cattle by mimicking the behavior of native herbivores. He visited ranchers in Wyoming, New Mexico, and Arizona who were bringing back endangered species from the edge of extinction. He visited others who were restoring devastated riparian areas as part of their livestock operations.

He called it a "new environmentalism" and took these stories back on the road as part of a lecture series, talking to diverse groups that included county commissioners, vegans, radical environmentalists, private property rights activists, agency representatives, and many more. The response, he said, was overwhelmingly positive. People were united on the goals that he described. "What naysaying I did encounter," Dan recalled, "came from the 'captains of conflict,' the leaders who have made a business of the hundred-year war between environmentalists and ranchers—the ones who have something to lose if the rest of us declare peace."

To illustrate the point about ends and means, Dan told his audiences the story about a meeting of the 6-6 group on a ranch near Tucson, Arizona, where the host rancher told an antigrazing activist: "Tell me what you'd like this place to look like and I'll make that my goal and work toward it. Then we can be allies instead of adversaries."

The antigrazing activist responded, according to Dan, by saying:

"There's only one thing you can do to make this place better. You can leave. Because if you stay, no matter what you do to the land, no matter how good you make it look, it will be unnatural and therefore bad. And if you leave, whatever happens to this place, even if it becomes as bare as a parking lot, it will be natural and therefore good."

To the activist, the land's health was prescriptive, not descriptive. Saying that it was all right for the land to become as "bare as a parking lot," even if it meant destabilizing the ecological function of the land, as long as it did not have any cows on it, is an example of what Dan calls "goals that incite." Consider most conflicts around the world, he says. Aren't they fundamentally about prescriptions: how to behave, how to worship, how to help the poor, fix an economy, settle a dispute. Want a healthy riparian area? Get the cows out. See, wasn't that easy?

Dan is more interested in "goals that unite": descriptive goals toward which we can all aim together. Although you might not be able to get Democrats and Republicans to agree on a political course of action, he says as an example, you could get them to work together to build a house. Or fix a creek. "Set a goal," Dan said, "just about any kind of goal, and humans will accept it as a challenge and work to achieve it if the result is something that matters to them, and healthy environments matter to just about all of us."

The discomfiting of his fellow environmentalists continued in 2005 with the publication of Dan's next book *The Gardeners of Eden: Rediscovering Our Importance to Nature*. In it he argues that the "leave-nature-alone" philosophy of the environmental movement over the last eighty years or so has done more harm than good and should be replaced by a new paradigm.

Dan believes we have become "space aliens" on our own planet. Once upon a time, humans enjoyed a mutualistic relationship with nature. In much the way that bees depend on flowers, beavers on creeks, and wolves on elk, many ecosystems evolved in the presence of humans, and over time their health began to depend on our species—to set fires, apply hunting pressure, and cultivate the soil. We were gardeners in Eden— natives living with, and using, nature symbiotically, though not always sustainably (as demonstrated by various megafaunal extinctions).

No more. Today, Dan writes, for the most part we "get our food, fiber, and other products from nature via a system of extractive technologies more characteristic of aliens than of a mutually interdependent community of natives."

Our transformation into an alien species of nature has had serious consequences: global warming, endangerment of plants and animals, erosion of the connection between ourselves and nature, and increasing isolation as more and more of us live in cities, surrounded by what Dan calls an "exploitosphere," where we extract everything from food to recreation.

One response to this transformation, more than a century ago, was the creation of a conservation movement, which aimed at correcting the exploitation of the natural world by both shielding places we valued from the destructive attentions of industry and by scientifically guided, "wise use" of natural resources, and by the rise of environmental advocacy, which emphasized legislation, litigation, and political activism. These efforts had at their core a fundamental purpose: mitigating environmental damage done by people. As we move deeper into the twenty-first century, however, Dan thinks these movements have led us to a dead end.

"Ironically, those countermeasures have been, for the most part, just as alien as the situation they were created to correct," he writes. For example, of the preservation wing of the movement he says: "We have created ever larger preserves and protected areas, and removed ourselves and our impacts from them. Acting as if we're trying to fool nature into thinking that we're not here, we have behaved as aliens would. We treat this land outside our exploitosphere as if it were a combination art exhibit, zoo, cathedral, and adventure park."

We play in nature—run, hike, ride, drive, raft, photograph, etc.—and then we go home to a de-natured universe of suburbs, highways, and strip malls. In the process, we've become detached from a meaningful relationship with nature built around work as "gardeners"—an ancient relationship that was founded on the belief that we are part of nature, not separate from it, as we have come to believe in recent decades. Therefore, our efforts at the protection and restoration of the natural world are doomed, Dan believes, because we've removed the key piece of a very complex puzzle—us. The answer, he says, is to become gardeners again, to recreate a more mutualistic relationship with nature through work.

At the core of our troubles, Dan says, is what he calls society's leave-it-alone approach to nature, which he first came to understand in those early 6-6 meetings, when his fellow activists argued that the only way to really heal the land was to protect it from humans. This leave-it-alone philosophy was a prescriptive approach with alienating consequences. "It is woven into the very essence of our society's awareness of what we call the 'environment,'" he asserts. "We think of it as a matter of fact, like gravity."

The ecological health of the ranches he visited, however, belied the leave-it-alone philosophy of his contemporaries, leading Dan to the conclusion that most people think the health of a piece of land is not a matter of its condition, but purely a matter of how that land is managed. "The leave-it-alone assumption," Dan says, "has brought us to the absurdity that the actual condition of a piece of land is irrelevant to determining if it is healthy or not."

The absurdity became painful to Dan as he began to study places where "no use" (read: no ecological disturbance) intersected with declining land health, such as the famous Drake Exclosure, in central Arizona, from which livestock had been excluded for forty years. The land inside the fence had become a biological wasteland. In contrast, the conditions beyond the fence, on grazed land, were healthier. Provocatively, Dan now calls the "protection" of land a type of "abandonment." What he seeks instead, and details in his book, are examples of nature/human mutualism and symbiosis. The key to the future, he believes, is maintaining the relationships that create life.

To demonstrate that Dan is not alone in his rather radical thinking on this topic, I'll cite two other examples. The first is an article titled "The Gardenification of Wildland Nature and the Human Footprint," by respected conservation biologist Daniel Janzen, published in the journal *Science* (vol. 279, 1998), in which he argues that there is no such thing as an "untrammeled" world anymore, anyplace. Not only is every block of wild country severely impacted by humans, we shouldn't even call them "wild" any longer. That's because they're not; they're gardens, albeit unruly ones.

According to Janzen, the whole planet is now a garden—under varying degrees of cultivation. "Gardens are beehives and cows," he writes, "and 16 varieties of rice growing in one rainforest clearing. Gardens are hydroponic tomatoes and vats of whiskey-spewing yeast. Kids do it, agroindustry does it, grandparents do it, astronauts do it, and Pleistocene Rhinelanders did it. And we will still be doing it 10,000 years from now."

As for wild places, Janzen says "Let's stop talking about national parks, wildlife refuges, conserved wildlands, biological reserves, protected areas, royal hunting reserves, national monuments, and all the other obfuscating labels that have been applied. Let's call them all what they are, wildland gardens."

And what does a wildland garden grow? The wilds. Ecosystem services. Biological diversity. The key to sustainable wildland gardening is work—restoration, succession, remediation, fire control, crowd control, biological control, and much more. But it's tricky, he notes— science and society can be difficult partners. "In the best of worlds we may achieve a very fine and finely negotiated partnership," he writes, "and in the worst of worlds, annihilation of one by the other. A wildland garden with gentle trodding from caring gardeners just might achieve the partnership. A wilderness faces certain annihilation as a battlefield."

The second example is from the book *Second Nature* by journalist and gardener Michael Pollan, who wonders if the garden could be a source of a new environmental ethic. He makes a list of possibilities:

- An ethic based on the garden would give local answers, including different solutions in different places at different times.
- A garden ethic would be frankly anthropocentric. He writes: "Every one of our various metaphors for nature—wilderness, ecosystem, Gaia, resource, wasteland—is already a kind of garden, an indissoluble mixture of our culture and whatever it is that's really out there."

- The gardener is not romantic about nature. He also has a legitimate quarrel with nature—"with her weeds and storms and plagues, her rot and death. Civilization itself is a product of this quarrel."
- The gardener accepts contingency. He learns to accept the hand he's dealt.
- The gardener knows she is not outside of nature. Not only is the "environment" alive and changing all the time in response to innumerable contingencies, one of these is the presence of the gardener him or herself.
- The good gardener commonly borrows methods, if not goals, from nature. "We do best in nature when we imitate her," writes Pollan, "when we learn to think like running water, or a carrot, an aphid, a pine forest, or a compost pile."
- If nature is one necessary source of instruction, culture is the other. In fact, our fate depends on our culture.

Both examples, along with Dan's provocative thesis, illustrate the advantages of a paradigmatic shift from the "leave-it-alone" school of environmental thought (and action) to an ethic based on land health, work, and collaboration. It doesn't mean tossing "the wild" out on its ear, but it does mean reentering the Garden of Eden again, and finding the wildness within.

Does Dan expect people to start chasing bison over cliffs again? No, but he does think we can start exploring goals that bring us together. This means a new approach not only for ranchers and environmentalists, but many members of society as well. Goals that incite, he reminds us, are all around us, wreaking their havoc. Goals that unite, in contrast, are just being discovered.

Washington, D.C.

Why would a national environmental organization with 500,000 members, a staff of 250, and a storied history of successful litigation turn to collaboration to achieve its goals? Because it works.

"For many years our unofficial motto was 'Sue the Bastards,'" said

Michael Bean, a lawyer and chair of the wildlife program for Environmental Defense (ED), a group that pioneered the lawsuit as a tool of environmental protection back in the late 1960s. "Today our official motto is 'Finding the Ways that Work.' It reflects an increasing pragmatism within the organization."

One approach that works is collaboration—helping private landowners, for example, reach economic and environmental goals while accomplishing ED's goal of protecting endangered species. This stands in stark contrast with ED's original approach: go to court.

This approach is traced back to Rachel Carson's *Silent Spring*, which alerted the world to the devastating impact of pesticides on the health of human and wildlife populations. In the mid-1960s, inspired by Carson's book, a group of scientists concerned over the use of the pesticide DDT on Long Island, New York, decided on a then novel strategy to combat environmental degradation: they hired a lawyer. DDT caused the eggshells of osprey, peregrine falcons, bald eagles, and other wild birds to become thin and break, endangering their survival. The pesticide persisted in the food chain as well, threatening human health. So the scientists went to court on behalf of the environment—and lost.

They knew the fight had just begun, however.

The group incorporated as the Environmental Defense Fund in 1967 and went on to win the war in 1972, when the federal government banned DDT throughout the United States, giving birth in the process to the profession of modern environmental law. The lawsuit was here to stay. Over the next two decades, however, ED also hired economists, engineers, and scientists in an effort to find an incentive-based approach to environmental protection that relied on economics as well as litigation, becoming, in the process, one of the nation's most effective and influential environmental advocacy organizations.

Part of their effectiveness, however, was due to their realization that lawsuits had limited utility. They were great for stopping the bad behavior, they learned, but not very useful for encouraging good deeds. "The lawsuit is a great hammer if every problem is a nail," said Michael Bean. "However, we've come to realize that lasting solutions to environmental problems require a different approach. A lawsuit, for example, isn't very useful for fixing global climate change. To do that you need to work cooperatively with people."

As an example, in 1995 Michael pioneered a strategy for the protection of endangered species on private land that is based on positive economic incentives as an alternative to confrontation. This strategy is called "Safe Harbor," for the image of a safe place to rest. Traditionally, the approach to the protection, restoration, and maintenance of endangered species in the United States was through the employment of the long arm of the Endangered Species Act, which became law in 1973 and is administered by the U.S. Fish & Wildlife Service. This law entails a bevy of rules and regulations, procedures, consultations, and compliances, any one of which could trigger the "hammer" of sanction by the government (usually as a result of a lawsuit brought by an environmental group), often in the form of restricted use of land.

The presence of an endangered species might, for instance, restrict a private landowner's right to develop the property for commercial use if that use damaged or reduced the habitat required by the species to remain viable. On public land, poor management of livestock can lead to damaged riparian areas in creeks, which can, in turn, reduce bird or fish habitat through increased erosion or reduced vegetative cover. In these situations, the Fish & Wildlife Service or another federal agency could restrict the use of the area by cattle, for example, or otherwise enforce a regulatory approach to habitat protection, often over the objections of ranchers, loggers, and others.

Working with the U.S. Fish & Wildlife Service (instead of Congress), Michael and others created a legal document by which landowners would be shielded from regulatory action if they undertook voluntary steps to assist a species in crisis. The idea was to encourage people to do good deeds for wildlife by protecting them from punishment.

For example, suppose a baseline wildlife survey determined that five acres of suitable habitat on a particular property was occupied by an endangered, federally listed species—the presence of which might otherwise restrict the landowner's use of the property. The landowner could enter into a formal agreement with the Fish & Wildlife Service that said if he or she created or restored additional habitat for the species over time, he or she would not be subjected to a corresponding increase in federal regulation.

When Safe Harbor came into existence officially, many environmental organizations greeted it with skepticism. Over time, resistance has dwindled, and for a straightforward reason: the carrot works. To date, more than three hundred landowners have Safe Harbor agreements in place, representing more than 4 million acres of land across the nation, protecting dozens of imperiled species.

In 2004, ED launched another conservation campaign aimed at endangered species and private land. Called "Back from the Brink," its goal is to encourage landowners to explore a wide range of incentive-based tools, many of them provided by the federal government through existing programs, such as the Farm Bill, which has a major section called the Conservation Security Program that pays landowners for agricultural practices that restore or maintain wildlife habitat.

"The idea is to reward private landowners for taking on public responsibilities," said Tim Sullivan, formerly ED's Rocky Mountain director. "That means getting ranchers and farmers involved in programs that literally put money in their pockets for activities that benefit endangered species."

Whereas the Safe Harbor program is opportunistic in a sense, focusing on individual landowners, the Brink campaign focuses on the species themselves—fifteen, to be specific, across the nation, including the northern aplomado falcon and the southwestern willow flycatcher—and the opportunities in federal agricultural programs to assist them. "The Farm Bill has huge potential to help farmers and ranchers achieve conservation goals," said Sullivan, "but many aren't aware of its benefits. That's what we are trying to do with this campaign—make people aware of their options."

Getting federal agricultural policy and programs to do conservation is something new under the sun, said Sullivan. In the past, environmental concerns were dealt with by environmental laws, such as the Endangered Species Act, or by a program like Safe Harbor, while agricultural programs were considered, well, for agriculture. Today, the effort is to steer policies that affect farming and ranching practices and environmental concerns toward each other. "We're trying to divert more public money into conservation while lowering the temperature overall on endangered species," said Sullivan. "That's why we've made such a big investment in the Farm Bill. We feel it has the potential to have a huge impact on conservation in this country."

As tools of conservation, Safe Harbor and Back from the Brink demonstrate what can be achieved when the incentives are voluntary, positive, and results-based. They are also examples of goals that unite, rather than incite.

ED hasn't laid down the tool of litigation, of course, preferring to use it these days for specific purposes, such as cleaning up dirty air. And these days much of ED's work is focused on global warming, which sometimes requires them to play rough; a recent campaign, for example, against a utility company in Texas led to a landmark deal that blocked the construction of many new coal plants. And they've worked hard to reduce emissions from automobiles by both legislative and litigious means.

Just as Dan Dagget doesn't fit the stereotype of a traditional environmentalist anymore, ED may not fit the stereotype of a traditional national environmental organization. Still, its support of the collaborative process demonstrates how diverse the movement has become in a short period of time. And ED's commitment to cooperative ventures with private landowners also demonstrates the chief attraction of collaboration: it unites.

Chapter Eight

QUID PRO QUO

Valle Grande Grassbank *Near Pecos, New Mexico*

In 1997, author and conservation leader Bill deBuys had an idea—and a problem.

The idea was a novel one: to get natural fire back on the land using the tool of a "grassbank," to be located on the Valle Grande allotment, a thirty-six-thousand-acre stretch of national forest land on top of Rowe Mesa, forty miles east of Santa Fe, New Mexico.

A grassbank is a physical place, as well as a voluntary collaborative process, where forage is exchanged for one or more conservation benefits on neighboring or associated lands. Generally, it is an empty (of cattle) stretch of land, publicly or privately owned, that is offered for livestock use in exchange for a conservation activity, such as rangeland or forest restoration, on the home ranch. Cattle come to the grassbank from the home ranch or allotment for a period of time sufficient to complete the conservation work. In other words, a grassbank is a quid pro quo—an exchange of grass for a tangible conservation benefit.

Inspired by a pilot grassbank on the privately owned Gray Ranch in southwestern New Mexico (the term "grassbank" was coined by rancher and poet Drum Hadley), Bill convinced the Conservation Fund, a national environmental organization, to purchase 240 acres of deeded land on top of Rowe Mesa. The property came with a year-round federal grazing permit but no cattle. Instead of buying cattle, Bill proposed to offer the grass of the Valle Grande allotment as a "bank" to national forest permittees around the region in exchange for restoration work on their home ground—principally forest thinning and prescribed fire. His partners included the U.S. Forest Service, the Northern New Mexico Stockmen's Association, and the Cooperative Extension Service, a division of New Mexico State University.

That was the idea—the problem was it almost never got off the ground.

That's because the key to the grassbank concept was another evolving, and controversial, idea at the time: collaboration.

The collaborative movement, sometimes called the watershed model because many community-based collaboratives (CBCs) are defined by watershed boundaries, had grown fitfully since the mid-1990s. CBCs generally involve diverse stakeholders in a consensus-based approach to long-term management of natural resources. The emphasis of these groups is on problem solving and the exploration of shared goals and concerns. Their success ultimately depends on their ability to foster strong *relationships* among the stakeholders over time. This contrasts with the advocacy model, in which groups with an interest in a particular natural resource struggle *against* one another for dominance.

Although many of the early CBCs, such as the Applegate Partnership in southern Oregon and the Quincy Library Group in northern California, enjoyed varying degrees of success, they definitely generated strong emotions, including a well-publicized broadside from Michael McCloskey, a national Sierra Club leader. In a memo to the board of directors of the Sierra Club in 1996, he wrote: "A new dogma is emerging to challenge us. It embodies the proposition that the best way for the public to determine how to manage its interests in the environment is through collaboration among stakeholders, not through normal governmental processes."

McCloskey identified what he considered serious flaws in the CBC idea, including

- the potential overwhelming influence of "industry" at the local level;
- the lack of strong environmental advocates in rural communities;
- the disempowering of national constituencies by transferring decision making to the watershed level;
- the "delegitimization" of conflict as a way of protecting natural resources; and
- the tendency, in his opinion, for CBCs to produce "lowest common denominator" results.

Advocates of CBC efforts countered that the process

- is more creative and produces comprehensive solutions to complex problems, resulting in better management on the ground;
- builds trust among diverse interests, which is necessary for progress;
- considers economic vitality and ecosystem health to be compatible;
- encourages the dissemination of knowledge across stakeholder boundaries so that all may learn;
- avoids costly and divisive litigation; and
- empowers rural residents who otherwise are often shut out of national-level policy and rule making.

Both sides make good points, in my opinion. At the time, however, the main issue was *power:* who had it, who had to give it up, who controlled whom, and so forth. Having emerged relatively victorious from the timber wars of the 1980s in the Pacific Northwest, and gearing up to tackle livestock grazing next, the environmental organizations active in the West were not keen on seeing their power undermined by CBCs. Simultaneously, many local community leaders, even though they were largely defeated, still preferred to brawl with environmentalists and government agencies.

In the middle were people like Bill deBuys. Bookish, gracious, funny, thoughtful, and inspiringly articulate (he once described a public debate between pro- and antilogging advocates as "hallucinogenic"), Bill had the intellectual chops, organizational skills, and philanthropic contacts to steer a course into the radical center. So, when a local captain of conflict in the environmental community challenged Bill to "do something" rather than keep complaining about all the lawsuits the activist was filing, Bill accepted the dare. That something was the Valle Grande Grassbank.

But it almost didn't happen. As he began to put the pieces of the grassbank together in 1997, opponents lined up as if they were at a shooting gallery. Forest Guardians, an environmental group dedicated to ending public lands ranching, objected officially to the grassbank's creation, citing the usual litany of complaints against cattle. The New Mexico Cattlegrowers' Association protested too, expressing their concern that a national environmental group would be getting into the livestock business. Even the lieutenant governor of New Mexico opposed the concept and worked aggressively, according to Bill, behind the scenes to derail the project (he sided with the Cattlegrowers' objections). Many others were lukewarm to the idea, including the region's mainstream environmental organizations.

It was touch and go until the two largest newspapers in the state, the conservative *Albuquerque Journal* and the liberal *Santa Fe New Mexican,* editorialized in favor of the grassbank, which caused the political opposition to fade away.

Why was there so much initial opposition to an idea that seemed to benefit all sides? Part of the answer can be found in the standard "taking sides" over livestock grazing on public land—what author Dan Dagget calls the "Conflict Industry." Another part of the answer was resistance among traditional ranchers to the idea of conservationists getting into the livestock business. But some of the opposition was due to the novelty of the grassbank idea itself. It was new, it was collaborative, and it was proactive.

But what exactly was it trying to accomplish? For Bill, the quick answer was this: restoring environmental health meant getting fire back on the land.

"In a detailed study of a 250,000-acre area in northern New Mexico," Bill wrote in a summary of the grassbank's goals, "ecologist Craig Allen found that between 1935 and 1981 tree and shrub encroachment had reduced the grassy component of the area's ecological mosaic by 55%."

This decline was regionwide, and continuing. Bill noted that not only did this decline have important ecological consequences, by diminishing ecological diversity, for instance, but it also eroded the economic viability of ranching in places such as northern New Mexico, where small-scale Hispanic ranchers depend upon the use of public lands for the continuation of their livelihoods and traditions. Too many trees and not enough grass were a recipe for conflict.

"Consider the dynamics," Bill wrote. "A fixed number of cows (and an increasing population of elk) must draw subsistence from a grass resource that is declining faster than one percent per year. The cattle necessarily use remaining grasslands heavily and crowd into riparian areas."

"The social consequences of such a predicament are also clear," he continued. "Environmentalists blame ranchers for the damaging effects of grazing and press for the removal of cattle from public lands. Ranchers, meanwhile, fight to maintain their herds and blame environmentalists for a range of sins that includes intolerance and cultural insensitivity. Locked in their accustomed roles, both sides fail to address the long-term ecological changes that structure their conflict."

To Bill, and many others, restoring grassland diversity and productivity, which should reduce social conflict, meant restoring fire to its natural role on the land. But that's not all. Restoration also requires, in many instances, mechanical thinning of woody species to reduce the risk of creating a fire that's so hot that it might do considerable damage to the land's regenerative capacity.

"Let it be noted that the simple removal of cattle from public lands," wrote Bill, "as urged by a substantial number of environmentalists, will not restore environmental diversity and health, for it will not bring the keystone process of fire back into the landscape."

But a grassbank could make that possible. The Valle Grande Grassbank, for instance, took cattle from forest allotments around the region, where local grazing associations have permits to run their animals, for two to three years so that restoration work could take place free of potential conflict with livestock-related interests. Bill saw a grassbank as a powerful tool because it could integrate environmental and economic goals, operate in harmony with local social and cultural traditions, encourage shared ownership, and meet environmental justice concerns.

To better describe what he was trying to accomplish, Bill wrote out three overarching goals for the grassbank:

- To improve the ecological health of public grazing lands for the benefit of all creatures dependent on them—from juncos to jackrabbits and curlews to cowboys
- To strengthen the economic and environmental foundation of northern New Mexico's ranching tradition, which is arguably the oldest in the nation
- To show that ranchers, conservationists, and agency personnel can work together for the good of the land and the people who depend on it

"In the case of northern New Mexico, we believe that the best hope for ecologically sound, fire-wise stewardship of public land lies within the ranching community," Bill wrote. "If ranchers, working with environmentalists, become advocates for prescribed burns, wildfires, and related treatments, political leaders and public agencies will respond accordingly—to the lasting benefit of the land."

For a while, it worked.

In the first six full seasons of operation, the Valle Grande Grassbank took more than two thousand head of cattle from nine separate grazing associations across two national forests in northern New Mexico. Conservation projects accomplished included the following:

- Prescribed fire: 5,590 acres
- Hand-thinning ponderosa or mixed conifer forest: 4,020 acres
- Brush/tree removal: 550 acres
- Riparian fencing: 5 miles
- Road improvements: 25 miles
- Trail improvements: 35 miles
- Water developments: 6
- Wetland/Playa projects: 4

It was a promising start, but like many new ideas, no mater how innovative or collaborative, the "beta test" of the real world eventually caught up to the Valle Grande Grassbank. In fall 2004, The Quivira Coalition took over its operation from Bill and the Conservation Fund. We had been assisting for a year or so with its operation, and Bill considered us the logical heirs to the project. We agreed, though in short order, reality began to bite.

First, the modest conservation gains came to an end during the final three grazing seasons (2004–2006) when no restoration work was completed by the Forest Service on the "home" allotments of permittees. This occurred for a variety of reasons, including drought, National Environmental Policy Act hurdles, and budgetary constraints within the Forest Service. But it exposed a weakness in the model: relying on an overworked, understaffed federal agency for the conservation part of the grassbank quid pro quo could be risky.

Second, the funding ran out. The grassbank's $160,000 budget was entirely grant-funded, and when the grants dried up, as they did at the end of 2006, so did the project. This raised a big question: how can grassbanks "pay" for themselves? It became clear to us that relying on the fickle and increasingly competitive world of federal grants and private philanthropy is rarely an economically sustainable strategy.

Third, the long distances traveled by permittees to get to the grassbank became increasingly problematic as transportation costs rose over time (participants paid their own way to the grassbank). A number of permittees, in fact, dropped out for this reason.

By fall 2006, all of these challenges came together. Some were resolved relatively easily, such as reorienting the grassbank to serve local permittees, but others proved more difficult to overcome, such as the funding conundrum. In fact, we decided to shut down the grassbank temporarily while we created a new business model that addresses these challenges—what we are calling "Grassbank 2.0." We still believe that the quid pro quo at the heart of the grassbank model is critical, as are the original goals of the project, but like an early version of computer software, its implementation needs an upgrade.

Bill anticipated this need for flexibility, which he saw as one of the strengths of the grassbank model.

"Our goal is to be consistently and continually adaptive," wrote Bill.

"If the land is changing, so must we. Our fundamental challenge is shared equally by both the conservation and ranching communities: how to respond to the constant dynamism of the lands upon which we all depend."

Heart Mountain Grassbank *North of Cody, Wyoming*

When the Nature Conservancy (TNC) purchased in 1999 the scenic fifteen-thousand-acre Heart Mountain Ranch, located on the east side of Yellowstone National Park, it considered selling the six hundred acres of irrigated ground that came with the property. After all, what did a one million–member worldwide conservation nonprofit organization dedicated to preserving rare native plants and animals and the habitat they require want with irrigated agricultural ground?

It had bought the ranch, located north of fast-growing Cody, Wyoming, to protect important wildlife habitat from the threat of development. Everything from elk to peregrine falcons to the endangered sage grouse called the ranch home. The property also harbored four rare plant species of concern. As a bonus, Heart Mountain itself was something of a must-see among geologists. But none of these attributes involved the six hundred acres of former hayfields, so TNC's first inclination was to sell the irrigated ground to a local farmer or rancher, reaping both the financial benefits of such a transaction and the feeling of goodwill that would come from helping the agricultural community.

But then Bruce Runnels and Laura Bell, both leaders in TNC, struck on an idea: why not create a grassbank?

A few years earlier, TNC would undoubtedly have sold those acres. But a lot had changed for the organization since its founding in 1946, when a small group of scientists formed the Ecologists Union with the goal of saving critically threatened natural areas, especially those with imperiled native species. Up until that time, the protection of wild and biologically diverse parcels of land had been primarily the job of the federal government, state agencies, or private hunting and game groups. Parks, forests, refuges, wilderness areas, and hunting preserves were the dominant means by which protection was provided. But was it enough? Some scientists thought not, and when the Ecologists Union changed its name to the Nature Conservancy in 1951, it embarked on what was then a novel strategy for preserving biodiversity: land acquisition.

It made its first purchase, sixty acres along the New York/Connecticut border, in 1955. Six years later it donated its first conservation easement, on six acres of salt marsh, again in Connecticut. By 1974, TNC was working in all fifty states, often in partnership with state and federal agencies. In 2000, it launched its Last Great Places campaign for land acquisition, education, and research. By 2007, TNC protected more than 117 million acres of land and five thousand miles of rivers and operated more than one hundred conservation projects in marine environments.

A turning point for the organization occurred in 1990, when TNC purchased the beautiful and biologically rich, 322,000-acre Gray Ranch, located in the bootheel of southwestern New Mexico. Sheltering more than 700 species of plants, 75 species of mammals, 50 species of reptiles, and 170 species of breeding birds, the Gray was considered one of the most significant ecological landscapes in North America in private ownership. For this reason, TNC purchased the ranch from its wealthy Mexican owner, who at one time had owned the Buenos Aires Ranch in Southern Arizona (which later became a national wildlife refuge), at a very high price, with every intention of getting it into the hands of the federal government as a wildlife refuge. Buying land and transferring it to the federal government was a popular strategy with TNC at the time, as well as with its dues-paying members.

It never happened. For starters, the Gray was also known as the Diamond A Ranch, one of the area's legendary ranching outfits, and thus was culturally important to the residents in the area. Second, it was still a working ranch, and thus a tax-paying, cowboy-hiring member of the local economy. For these and other reasons, the rural community surrounding the ranch went nuts when they heard news of the purchase. They complained loudly to TNC officials, elected representatives, newspaper reporters, and anyone else who would listen that enough was enough. This process of buying ranches and turning them over to the federal government for "protection"—à la the Buenos Aires National Wildlife Refuge—was not acceptable to them.

To their surprise, they found sympathetic ears at TNC, which was hearing similar complaints in other places. So the organization asked itself an important question: could it accomplish its conservation goals and keep the ranch in private hands as a working cattle operation? And perhaps just as importantly, could it find a "conservation buyer" who would help them recoup their very substantial financial investment in the property?

The answer to both questions proved to be "yes." In 1993, TNC sold the Gray to Drum Hadley, a local rancher who also happened to be an heir to the Budweiser beer fortune. After the sale, Drum and members of his family created the Animas Foundation, named for the nearest town, to protect the property and to ranch it for conservation as well as community goals. A year later, they became charter members of the newly minted Malpai Borderlands Group—and what had once been a very contentious landscape transformed quickly into a model of collaboration and conservation.

Fast forward to 2005. TNC now manages two grassbanks, one at the Heart Mountain Ranch, in Wyoming, and the other on its sixty-thousand-acre Matador Ranch, in northeastern Montana. Both have roots in TNC's experience with the Gray Ranch, in its quest for new strategies that achieve conservation goals. That's why they decided ultimately to keep the six hundred acres of highly productive ground at Heart Mountain to use as a grassbank, so they could leverage its forage for biodiversity protection and habitat restoration.

By 2002, the Heart Mountain Grassbank was enjoying its first full season of operation. It was also enjoying strong local support from the ranching community, as well as full cooperation from the Forest Service, the U.S. Bureau of Land Management (BLM), and the state Game and Fish Department. In 2004, the last major piece of the puzzle fell into place when Maria Sonett and her partner, Skip Eastman, were hired to direct the grassbank and manage the ranch.

Maria came to Heart Mountain from the Valle Grande Grassbank, where she had worked as assistant director for Bill deBuys. It was an unexpected turn in the life of an individual who had become infatuated with grass. Born in northern California, Maria moved to Tucson, Arizona, as a teenager when her father became a professor at the university. Maria had no initial interest in agriculture; but she did love horses. "I started riding when I was five," she recalled. "We lived on the edge of ranches in California—maybe not what we would call a ranch today, but they all had horses. And I thought it was the greatest thrill to go out and catch a neighbor's loose horse."

She was allowed to keep one of the horses she caught, and she has been involved with horses ever since. As a teenager, when her friends were babysitting, Maria was riding other people's horses for money.

As a child of the suburbs, Maria admits she "didn't pay much attention to what was under the hooves of my horses." It wasn't until graduate school that she began to develop what has become a lifelong love affair with grass and grasslands. It all began with a move to New Mexico.

"One day I got off my horse and decided to collect all the range grasses that I didn't recognize, which was most of them," she said. "I was in grad school at UNM in landscape architecture at the time focusing on the 'hardscape' of parking lots and handicap spaces. I hated it. I wanted to learn how to read the natural landscape instead."

She took the grasses down to the Albuquerque office of the Soil Conservation Service (now the Natural Resources Conservation Service), where she met John Warner, who became a mentor. She volunteered in the office, becoming increasingly fascinated by grass, grasslands, and the range issues entangled with both. "Grass is a lot like an ocean," she commented, "There's an inherent mystery in it."

In the meantime, she was raising two kids and worrying about money. With the encouragement of Professor Bill Fleming at UNM, she abandoned hardscapes and wrote her master's thesis instead on mine reclamation. "How to grow grass on impossible soils," is how she described her research. "It's part of what they call restoration ecology today. In any case, it led to a job with a private firm, thankfully."

Two years later, she found herself directing the Heart Mountain project. "I hate the term 'win-win,' but it fits the grassbank idea," she said of her work. "It reduces conflict because it gets conservation done and it helps the ranchers."

According to a TNC publication, in 2004 the Heart Mountain Grassbank received 365 cow-calf pairs from the Sage Creek Grazing Association, northeast of Cody, which enabled the BLM to carry out 3,600 acres of sage grouse habitat improvement using prescribed fire and mechanical treatment. In the two previous years, the grassbank took cattle from the nearby Shoshone National Forest, which enabled the Forest Service to burn hundreds of acres for Douglas fir forest for habitat improvement adjacent to the North Absaroka Wilderness area.

The goal of the BLM projects was to improve habitat for the imperiled sage grouse, populations of which have declined by 40–80 percent throughout their historical range in the West. The current population is estimated to be 140,000—about 8 percent of the historic population. Its low numbers are principally due to the spread of sagebrush, which in its mature form can become a sort of monoculture across landscapes, due to the absence of fire. The BLM mowed hundreds of acres of sagebrush on the temporarily vacant (of cattle) Sage Creek allotment and burned an additional 120 acres, which created a patchy mosaic among the plants, a pattern preferred by the grouse.

"We couldn't have done this [the treatments] without the grassbank," wrote Tricia Hatle, BLM range conservationist for Sage Creek. "We have to rest the range for a couple of seasons while we do the project. Grasses need to get built up in order to carry a fire, and then the vegetation needs a chance to regenerate after both the fire and the mechanical treatment."

Ranchers benefit too. Each pays $15 per season for each cow-calf pair to graze on the six hundred acres of irrigated pasture, which is set up for high-intensity/short-duration grazing, and each animal can gain two to three pounds a day.

"We've installed new filter systems, buried pipe, built new fences, repaired lots of old ones, monitored utilization and regrowth on our irrigated ground, experimented with state-of-the-art organic fertilizer, and planned many more projects," said Maria. "But one of the more difficult tasks we've had is not running over the bumper crop of sage grouse, who move to the irrigated ground to grow up. Wherever we walk, ride, or drive, we send numerous young grouse scrambling into the air."

In 2005, Maria, Skip, and TNC expanded their efforts by putting a herd of cattle on the unirrigated ranch lands. They were cows from a Forest Service allotment in bighorn sheep habitat. TNC and the ranchers have made a ten-year commitment to each other.

"It was a happy confluence of needs," said Maria.

Whether the Heart Mountain Grassbank has a happy ending remains to be seen, however. By fall 2007, both Maria and Linda Poole, the director of the Matador Ranch Grassbank in Montana, had resigned from TNC and moved on to new jobs with other organizations. Though TNC intends to keep both grassbanks going, their future is unclear at this point in time, despite their success. Neither was "profitable" in the traditional sense—echoing the financial instability that closed down the Valle Grande Grassbank—requiring TNC's deep pockets to keep them operating. Indeed, money may be the principal reason why, more than ten years after Bill deBuys launched his novel idea, there are still only three formal grassbanks across the whole West. Collaboration set the foundation for an innovative approach to land stewardship, of which grassbanks are a good example, but money is still the mother's milk of long-term success.

In fact, the future of grassbanks likely depends on developing what some of us have begun to call "conservation with a business plan"— figuring out how to do conservation entrepreneurially so that it can reduce its dependence on subsidy, whether that subsidy comes in the form of government grants, private philanthropy, or deep pockets. But we're not there yet. In the meantime, there's much to be learned from collaborative efforts in all their various shapes and sizes. They face many challenges, and certainly don't answer every conservation concern on the land, but they're hopeful because they work on the positive side of the ledger—on the side that says relationships matter, handshakes matter, health matters, and ideas matter.

Which, after all, is what quid pro quos are all about.

A MACHO FAILURE

Macho Creek *North of Deming, New Mexico*

For years, the Macho Creek demonstration project has been a staple of my slide show, invariably following the "opening credits" on the mission of The Quivira Coalition and the New Ranch. If I have time I'll expand on the coalition's origins, the ideas of Allan Savory, and the search for neutral ground in the grazing debate. Then it's on to Macho Creek.

I chose to talk about the Macho Creek project for four reasons: first, it was our earliest demonstration project; second, it is a simple and effective example of the New Ranch toolbox in action—the benefit of dormant season grazing in this case; third, the dramatic before and after photos always take the audiences' breath away; and fourth, well. . . .

The fourth reason was the project's success—a "win-win-win" for land health, ranching, and public land management. It was collaborative, voluntary, science-based, innovative, educational, proactive, and provocative.

And when the clock struck midnight the project turned into a pumpkin.

In May 1998, The Quivira Coalition helped the New Mexico State Land Office and Quail Unlimited build a 2.5-mile-long electric fence along the western edge of a riparian zone on state land along Macho

Creek, located north of Deming, New Mexico. The goal was a simple one: to create a riparian pasture to keep the cattle out of the creek during the growing season.

The reason to do so was also straightforward: the riparian area had been overgrazed to the point of being largely nonfunctional. By that, I explained, it lacked riparian vegetarian—rushes, sedges—the absence of which can cause accelerated erosion, lowering the stream channel, which can drop the water table . . . You get the picture.

But we don't like to use the word "overgrazed" in public discourse. One of our board members, a public lands manager himself, put it more tactfully, calling it a "management opportunity." Our idea was to build the fence, let the cows graze in the creek only during the dormant, or winter, season and monitor the ecological response.

It was an attractive project on many levels, not the least of which—and this is one of the points I make in the slide show—was how it rebuffed the position of antigrazing activists that riparian recovery and cattle grazing are mutually exclusive. By encouraging ecological recovery while the land is *simultaneously* used for agriculture, we hoped to demonstrate that the debate over cattle in riparian areas need no longer be painted in black and white.

At the same time we hoped to demonstrate to the local rancher how a new tool—dormant season grazing—could help his bottom line. And we also wanted to see the land become healthy enough to again support game birds, which is one reason why hunters from Quail Unlimited were there stringing wire on a hot May day.

All went well at first. The "fence raising" proceeded smoothly, and the cows stayed out of the creek bottom that summer. We contracted with Hawks Aloft, a nonprofit organization based in Albuquerque, to conduct five-times-a-year bird surveys in the riparian zone, and the U.S. Department of Agriculture's Jornada Experimental Range agreed to do quantitative monitoring of the vegetative response.

And what a response there was! When I returned to Macho Creek in September 2000, I could hardly believe my eyes. There was grass everywhere, including a species that Jim Winder, who ranched nearby, said he had not seen in a long time. The intervening years had not been particularly wet ones either, which is indicative of how quickly riparian areas can recover if "set free" from the effects of overgrazing.

The data from Hawks Aloft's surveys backed up our impression. Gail Garber, who directs the nonprofit, reported that winter surveys had shown a steady increase in bird species over three years. In particular, a consistent increase in blue grosbeak and Bullock's oriole specifically indicated improvement in understory and overstory vegetation along the creek.

I experienced my own "indicator" of riparian recovery while taking photographs that fine September day. While taking "after" shots of the recovery with my camera, a red pickup truck pulled off the road abruptly, disgorging a man armed with a shotgun who quickly raised the gun to his shoulder and began shooting. Unfortunately, he aimed in my general direction, so I quickly scooted out of there. Two years prior, needless to say, no one would have pulled off the road to shoot birds on this stretch of Macho.

I happily inserted the photographs into my talk, where they remain to this day, though not for the reason I originally intended.

The trouble began with the complex nature of the grazing arrangement. The permittee on the state lands was a college professor who lived in California. He had subleased his permit to a rancher who owned the neighboring land upstream. Alas, this rancher wasn't exactly an eager participant in the project. After all, it was his cattle that had created the "management opportunity" in the first place.

So, we had an absentee owner and a reluctant rancher as partners. Additionally, the small staff of the State Land Office was stretched far too thin over the 11 million acres under its jurisdiction. When Hawks Aloft began seeing trespass cattle in Macho Creek in summer 2001, we began to worry. The electric fence, which by its nature needs constant attention, was down, they reported, and in need of repairs.

What happened next, however, was the important lesson. The subleasing neighbor sold his ranch and moved away. In other words, the on-the-ground steward, no matter how reluctant, was suddenly gone. It got worse. The new owner of the upstream ranch, as it turned out, was not interested in cattle and declined to become the sublessee. The college professor was not able to do much long distance; and the State Land Office tried, but failed, to find a new rancher to manage the land.

Meanwhile, trespass cows were having a field day on all that grass we had grown. We didn't have enough money to hire a crew to fix the fence. Neither did the Land Office (almost all money the Land Office generates goes straight to public education). And the professor apparently just wasn't interested. It wasn't a happy situation.

In July 2003, I returned to Macho Creek for more photos. I braced myself. Driving up the road I spooked a herd of twenty-five cattle who clearly had made a home for themselves in the creek. I walked to the photo point where the hunter had pointed his gun in my direction nearly three years earlier. There were no birds, there was no grass.

No hunter would be returning to Macho Creek anytime soon.

The electric fence lay at my feet in a tangled mess, hopelessly beyond repair. And tangled with it was a lesson about the value of stewardship. We lost the steward of Macho Creek when the rancher moved away. He was a reluctant steward, to be sure, but while he remained, the land grew healthy again. As soon as he was gone, and no one replaced him, the land deteriorated again.

A cynic might say this makes a case for no steward at all—that kicking the cows off the land is the only way to ensure ecological recovery (keeping them off, by the way, is easier said than done). The lesson learned in Macho Creek, they might say, is what happens when the cows came *back*.

The lesson I learned is different: our land needs more and better stewardship, not less. It needs the active manager on the ground, watching the indicators of land health, fixing fences, and moving cattle around. Our land needs more care and attention, not less—and less is what you get when people leave, live too far away, don't have the necessary resources, or, as in the case of the new neighbor on Macho, aren't interested.

Restoration is an active word—I tell audiences it needs human involvement, guidance, and maintenance. And it needs to be done by someone who feels affection for the land, who lives there, and who is an eager participant. The alternative is more of what got us into this situation in the first place.

Today, Macho Creek is a "management opportunity" again.

How typical is it of other landscapes in the West? Very, I think. The need for more and better stewardship, not less, is great all across the region. The challenge of finding the financial means to do the work, however, is just as great. So too are the challenges presented by changing ownership patterns in the West, with increasing quantities of private land transferring to owners who may not have an active interest in agricultural or conservation activity. And Macho is typical of the complexity we all face in restoring—and maintaining—land health.

Chapter Nine

THINKING LIKE
A CREEK

Many Riparian Areas *New Mexico and Arizona*

During one of my travels, I heard a story about a man who had put short fences across a cattle trail in the sandy bottom of a canyon in Navajo country. His goal, I was told, was to force the cattle to meander in an S-pattern as they walked, encouraging the water to meander too, and thus slow erosion. I thought this idea was wonderfully heretical. That's because the standard solutions proposed for cattle-caused erosion were: (i) kick the cows out (if you were an environmentalist); (ii) ignore it and hope the problem fixes itself (if you were a rancher); or (iii) spend a bunch of money on diesel-driven machines and other heavy-handedness (if you worked for an agency). Putting fences in the way of cattle and letting them do the work? How cool.

Not much later, I met the author of this heretical idea while attending an environmental restoration conference in downtown Phoenix, Arizona, which was an oxymoron of unsettling dimensions. Someone told me that Bill Zeedyk would be giving a talk. I pricked up my ears. You mean that guy who's been trying to keep part of Hubbell Trading Post from washing away by putting sticks and rocks in the nearby creek bed, I asked? The guy who refused to use cement, riprap, or rock-filled wire baskets? It was, actually, and I bumped into him in the hall. Bill is hard to miss—he looks like a Dutch version of Santa Claus, with ruddy cheeks, twinkling blue eyes, a squarish salt-and-pepper beard, and a modest roundness at the girth that completes his aura of avuncular charm. Only don't tease Bill too much, as I eventually learned. Bill takes what he does quite seriously.

And what he does is help creeks get better. That might sound an odd job description, but given the standard environmental saw that over-grazing has degraded 80 percent of the region's creeks and riparian areas (the "wet zone" on both sides of a creek), compromising their high ecological value in the arid Southwest, strategies of restoration had become an important issue economically, environmentally, and politically over the last decade (thus the conference in Phoenix).

After making quick introductions, I asked Bill if the story about the fences and the cattle trail in the canyon bottom was true. It was, he said. Recognizing that water running down a straight trail will cut a deeper and deeper incision in soft soil with each storm event, increasing the probability of serious erosion trouble, Bill talked the local Navajo ranchers into placing fences at intervals along the trail so that the cows would be forced to create a meander pattern in the soil precisely where Bill thought nature would do so if the trail were a stream. Water likes to meander—it's nature's way of dissipating energy—and it will do so again even if it's temporarily trapped in a cattle-caused rut (or human-caused hiking trail), though it might take a long time. His fence idea was a way to speed the process up, he said.

What happened after the fences were put it in, I asked? The water table came up as vegetation grew back, he replied, because the water was now traveling more slowly and had a chance as a result to percolate into the ground, rather than run off like before. Steep, eroded banks began to revegetate as the water table rose and more water appeared in the bottom of the canyon, which encouraged riparian plant growth.

"Nature did all the heavy lifting," he said, before adding a warm, knowing smile. "It worked too, until someone stole the fences."

I followed him to his talk. Bill's comment reminded me that environmental problems are, at bottom, "people problems." One is inextricably intertwined with the other. Fixing the environmental problem without addressing the people part, to paraphrase Aldo Leopold, is like fixing the pump without fixing the well.

However, at age seventy, Bill would rather leave the people problems to somebody else. "I'm done arguing," he said to me. "I'd rather focus my energy on fixing creeks and roads."

And that's exactly what he has been doing. Beginning in 1995, five years after his retirement as a biologist with the U.S. Forest Service, Bill has developed an important set of techniques designed to "heal nature with nature," as I heard in Phoenix that day.

In the presentation, he inventoried this toolbox, illustrating how his low-cost, low-tech methods reduce erosion and sedimentation, return riparian areas to a healthier functioning condition, and restore wet meadows and other wetlands, all at a minimal cost compared to other techniques, such as the backhoe/rock-and-wire gabion-structure approach used by many landowners across the nation.

Bill's toolbox includes:

- one-rock dams (small structures that are literally one rock high)
- picket baffles and deflectors (wedge-shaped structures that steer water flow)
- wicker-weirs (rows of sticks that create a riffle effect in creeks)
- vanes (a row of logs pointing upstream that deflect water away from eroding banks)
- headcut control structures (that slow or stop the relentless march of erosion up a creek)
- worm ditches (that redirect water away from headcuts in wet meadows)
- "Zuni" rock bowls (small structures that trap water so vegetation can grow)

Many of these structures are placed directly in a water course. Vanes and baffles, for instance, often constructed of wooden pickets (harvested locally), are used to deflect stream flow. Weirs are used to control streambed grade and pool depth. One-rock dams are used to stabilize bed elevation, modify slope gradient, retain moisture, and nurture vegetation.

The goal of these structures, I learned, is to stop downcutting in creeks and streams, often by inducing an incised stream to return to a "dynamically stable" channel through the power of small flood events. Bill calls it "induced meandering." Its goals are to restore channel dimensions, reestablish appropriate meander patterns and pool/riffle ratios, restore stream access to its floodplain, and raise the alluvial water table, which enables riparian vegetation to grow.

In other words, when a creek loses its riparian vegetation—grasses, sedges, rushes, willows, and other water-loving plants—to overgrazing by livestock, say, it tends to straighten out and cut downward because the speed of the water is now greater, causing the scouring power of sediment to increase. Over time (and sometimes not very much time), this downcutting results in the creek becoming entrenched below its original floodplain, which causes all sorts of ecological havoc, including a drop in the water table (bad for trees and wet meadows). Eventually, the creek will create a new floodplain at this lower level by remeandering itself, but that's a process that often takes many decades. Bill's idea is to goose the process along by forcing the creek to remeander itself via his vanes, baffles, and riffles, carefully calculated and placed. And once water begins to slow down, guess what begins to grow? Willows, sedges, and rushes!

"My aim is to armor eroded stream banks the old fashioned way," said Bill, "with green, growing plants, not with cement and rock gabions."

The employment of one-rock dams typifies Bill's naturalistic approach. The conventional response of landowners over the years to eroded, downcut streams and arroyos has been to build a check dam in the middle of the watercourse. The old idea was to trap sediment behind a dam, which would give vegetation a place to take root as moisture is captured and stored. The trouble is, check dams work against nature's long-term plans.

"All check dams, big or small, are doomed to fail," said Bill. "That's because nature has a lot more time than we do. As water does its work, especially during floods, the dam will be undercut and eventually collapse, sending all that sediment downstream and making things worse than if you did nothing at all."

"The trick is to think like a creek," he continued. "As someone once told me long ago, creeks don't like to be lakes, even tiny ones. Over time, they'll be creeks again."

One rock dams, by contrast, don't collapse—because they are only one rock high. Instead, they slow water down, capture sediment, store a bit of moisture and give vegetation a place to take root. It just takes more time to see the effect.

"As a species, we humans want immediate results. But nature often has the last word," said Bill. "It took 150 years to get the land into this condition; it's going to take at least as long to get it repaired." The key is to learn how to read the landscape—to become literate in the language of ecological health.

"All ecological change is a matter of process. I try to learn the process and let nature do the work," said Bill, "but you've got to understand the process, because if you don't, you can't fix the problem."

Why is this important? Why even worry about fixing creeks in the first place?

For starters, there is a good reason why many authors and historians, including Wallace Stegner, have labeled the American West the "plundered province." More than a century of very hard use, including massive overgrazing by millions of livestock during the boom years between 1880 and 1920, has created a legacy of damaged, degraded, and just plain worn out landscapes across the region. Add clear-cut logging, thousands of mines, hundreds of thousands of miles of badly designed and poorly maintained roads, extensive oil and gas exploration, and a thousand other injuries from the plundering behavior of individuals and corporations over the years, and you have a region that is chronically in need of a good doctor.

This is hard for most Americans to understand, I think, because we spend most of our recreation time in parks, wilderness areas, wildlife refuges, and other pretty places that either *seem* healthy, or, in fact, *are* relatively healthy. That was certainly my impression growing up in the West. Backpacking through one national park after another, even hiking through the desert as an archaeologist, I had no idea what "land sickness" looked like, to use Leopold's phrase, other than the obvious signs of abuse. That changed when I began to work with Jim Winder, Kirk Gadzia, and others, and sped up considerably when I saw the land health map of the Altar Valley, in southern Arizona. But it wasn't until I walked up a deeply eroded arroyo one fine, sunny day in 2003 that the magnitude of the problem struck me like a bolt of lightning.

It happened at the boundary between the Gila National Forest and Jim and Joy Williams' ranch, located a few miles south of Quemado, in famously cranky Catron County, in west-central New Mexico. We were there as a result of a project we were doing with Bill Zeedyk on Loco Creek, which is located on the Williams' ranch and is an ephemeral tributary of Largo Creek, a substantial watercourse in the area.

We had met Jim and Joy Williams in Pie Town, New Mexico, in June 1998 when I accepted an invitation to speak at a meeting organized by three local women who despaired over the social and economic cost that constant conflict between ranchers and environmentalists had brought to their communities. Jim and Joy despaired too, but for a different reason: the Williams Ranch was in trouble. In 1995, the Forest Service reviewed the Williams' grazing allotment and decided to cut the number of permitted cattle they could run on the forest. It was the first time the permit had been cut in Jim's lifetime; he was then in his early fifties. Moreover, it had never been cut during the lifetime of his father, Frank, who had assembled the ranch back in the 1940s. The issue of contention was the condition of the land, which the Forest Service insisted was being hit too hard by Jim's cattle.

It was a common story at the time, with a common outcome. Angered by what he thought was the Forest Service's intransigence, Jim joined a class-action lawsuit with other ranchers against the agency. He also closely tracked another court case, this one brought by environmentalists upset at the government over cattle grazing on public land. "I thought the only answer was to fight," Jim told me later. "Well, we lost both of those cases, and so I thought that was pretty much the end of everything."

Financially struggling, and with their up-and-down relationship with the Forest Service at an all-time low, the Williams family, the last full-time ranchers in the Quemado area, began to seriously contemplate the one option that remained: to accept the offer of a subdivider to buy their substantial private land. Unwilling to take this option just yet, however, Jim raised his hand at the end of the meeting in Pie Town and invited The Quivira Coalition for a tour of his ranch, which we organized two months later. Liking what he heard us say about land health, progressive ranch management, and collaboration, Jim invited us back for further discussions. He also ordered the Catron County manager, who was on the tour, off his land when he tried to talk Jim out of cooperating.

Working with John Pierson, the local Forest Service range conservationist, and Kirk Gadzia, Jim set new goals for the ranch and began to sketch out a new plan of cattle management. Using existing fences and natural boundaries, they divided the ranch into smaller pastures and planned rapid moves of cattle through them. Jim agreed to graze his Largo Creek pasture in the winter months instead of late spring, as he had traditionally done. Jim and Joy also agreed to let Hawks Aloft, a nonprofit group hired by The Quivira Coalition, do bird monitoring on Largo Creek on their private land (to document the creek's ecological improvement). Everything went well: the grazing rotation worked, the land improved, communication and trust between Jim and the Forest Service was restored. Jim even joined in on the bird surveys.

"I got a real kick out of looking for ferruginous hawks on my place," Jim told me, referring to an elusive and sensitive species "of concern." That's probably not something a Catron County rancher would have said in the mid-1990s.

An important fruit of this trust-building blossomed in 2001, when Jim and Joy opened their private land to The Quivira Coalition for a riparian restoration project along Largo Creek.

Not only was the creek in need of doctoring, the ranch met an important precondition for Bill Zeedyk in any restoration project that he undertook: the livestock grazing had to be under control. There is no point to armoring a stream bank with riparian vegetation if the cattle come in and eat it all to the nub. As a consequence, Bill avoids working with landowners who allow overgrazing, thereby creating a very important link between the New Ranch and riparian restoration. One reinforces the other—good cattle management helps the grass grow along the creek, and the riparian restoration can increase the amount of forage available for animals.

All of this involved a steep learning curve for me, but nothing quite prepared me for what happened when we turned our attention to Loco Creek, a tributary of the Largo. It was called "Loco" for a reason—it was crazy to look at. Parts of it were so deeply entrenched that the walls rose above my head as I walked up it. According to Jim, it wasn't even a creek; it was an old wagon road that had eroded so badly over time that it intercepted the watercourse, redirecting it entirely. And it was eroding so badly that with each major cloudburst its bottom would drop noticeably, kicking huge amounts of sediment into Largo Creek, which is not a good thing. In fact, it is precisely the sort of environmental trouble that has agencies such as the U.S. Environmental Protection Agency (which administers the Clean Water Act) worried across the region.

It wasn't the sediment that got my attention, it was something farther up Loco Creek. Walking up the drainage one day, as crews under Bill's direction were placing erosion control structures farther down, I came to the boundary between Jim's ranch and the national forest. Stretched across the creek and ten feet above my head was an old barbed wire fence, complete with fence posts. I knew from a conversation with Jim that the Forest Service had built the fence in 1935. As I stood and stared at the fence hanging uselessly in the air, I recalled a story about Aldo Leopold. In the 1920s, while touring the national forests in the Southwest as part of his duties to review their condition, he observed significant signs of erosion taking place, often along creeks and rivers. Putting two and two together, he speculated that the hydrological unraveling that he saw was connected to heavy livestock grazing, and he reported his concern to his superiors (who took little action). I was looking at what Leopold had warned about, many decades removed.

My god, I thought. I stared up at the fence for a while longer. Then I took a photo.

Later I asked a man who works for the Natural Resources Conservation Service, which is a branch of the U.S. Department of Agriculture that works with private landowners, how much of the rest of New Mexico existed in a degraded condition similar to what I saw in Loco. "Most of it," he replied.

Bill Zeedyk agreed. During the 1980s, while serving as the regional director of wildlife and fisheries for the Forest Service, a position he held for fourteen years, he had an opportunity much like Leopold's—to survey the condition of forest lands—and much of what he saw disturbed him. It motivated him too.

Actually, Bill's second career—his present one as a restoration specialist—began while he was working for the Forest Service and it had a simple genesis: bad roads. As he traveled, Bill began to notice the deleterious effects of roads on wet meadows. Whenever a road crossed a wet meadow or a stream, he observed, almost invariably the culvert was installed too deeply, often with profoundly adverse effects on adjacent landforms.

"The culvert created a headcut that subsequently drained the meadow and often destroyed the resource," said Bill. Headcuts are two- to three-foot–tall dry "waterfalls" in a meadow, often started by poorly placed culverts under roads, that move inexorably upslope, lowering the water table and increasing erosion. "That's because the only goal of road construction at the time was cheap and safe roads with no regard for offsite effects."

While on a turkey hunt in the Zuni Mountains in 1985, Bill had a revelation. Disappointed that he wasn't finding any turkeys, he paused while hiking back to his truck and suddenly took notice of a deep gully alongside the road. Following the gully, which started with the culvert under the road, he saw that it had drained a big meadow nearby. "It was like a light went on," he recalled. "Maybe that's why there weren't any turkeys—the road had destroyed a key part of the habitat."

Bill went back to his boss, the regional forester, and got his concurrence to organize a team, which included the regional engineer, a hydrologist, members of the range staff, and others. They took a field trip to the drained meadow in the Zunis. To his surprise, their reaction was positive—they agreed that roads needed more attention—which led Bill to consider the power of collaborative thinking in problem solving. More immediately, it led to new designs in road crossings on forest lands. He also set about engaging the wildlife conservation community, of which he was an active member, in meadow restoration projects.

In the process, he surveyed a lot of roads, which eventually led him to author a book on road repair and maintenance in 1995 (and another one in 2006).

As an illustration of the impact roads can have on water, as well as the troubling legacy of our land illiteracy, Bill likes to tell the story of a rancher he knows in the Estancia Basin, east of Albuquerque, who built a road around a wet meadow to his daughter's house. The rancher knew enough not to build straight across the meadow; so when the meadow dried up anyway, the rancher blamed the prairie dogs.

"But he did it himself," said Bill, "by starving the meadow of its water supply from the microwatersheds that surrounded it. What impressed me was that after some discussion, he recognized and acknowledged the problem."

The landowner was willing to try one of Bill's simple but unconventional remedies. Bill directed a backhoe operator to excavate "rolling dips" every three hundred feet or so in the road. This allowed the water to flow back across the road and into the meadow, instead of being captured in a ditch along the road.

The meadow came back to life. "I really respect him for getting at the real nature of the problem. Too often we treat water as a nuisance, not a resource," said Bill, who calls his method of road repair "water harvesting."

Like many restorationists, Bill likes to think in terms of opportunities, not obstacles. And he thinks we're missing many of them.

"A huge part of our erosion problem in the Southwest is a result of bad roads," said Bill. "This is equal to overgrazing, in my mind. In fact, I like to say that 2 + 2 = 5. Poor roads plus overgrazing by livestock is greater than the sum of the parts. That's because less grass means more water coming off the hillslopes, which gets captured in a gully created by a culvert or a bad road ditch, which increases the erosive power of the water and causes even more erosion. In the meantime, the water is lost from the land it might otherwise nurture."

"But you can't just fix the roads," he warned, "you must also do proper grazing. And vice versa."

How Bill came to his restoration career says as much about his generation, and how far it has come over the decades, as it does about him. Born in New Jersey to schoolteacher parents in 1935, in what was then a rural area, he attended the University of New Hampshire, where he majored in forestry, having decided at the tender age of fourteen that he wanted to be a forester. He liked to hunt, fish, and trap—in fact, he paid for his first year at college by trapping muskrats. This led to his interest in habitat management—he wanted to trap more muskrats.

"Trapping taught me how to observe wildlife and encouraged a sensitivity to habitat needs. It taught me how to read a landscape."

Despite his burgeoning respect for nature, however, Bill grew up in an era when humans assumed they knew best. "We were always looking for a better tool to control nature," he recalled. "That changed with Earth Day, when we began to see that there are consequences to all that we do. Up until then we rarely took responsibility for our actions."

These attitudinal shifts extended to Bill's employer. He joined the Forest Service in 1962, becoming the first biologist on the Daniel Boone (then Cumberland) National Forest in the mid-Appalachian Mountains. He believed firmly in the wisdom of multiple use on public lands (and still does) because of its inclusiveness.

"Everyone stood to gain something from the common management of our forests, and this made the public lands system strong," he said. "Unfortunately, today the interests are splintered and the support for public lands has eroded to the point where I believe their future may be in doubt. There is no longer the bond of common ownership that protected the integrity of the National Forest system."

As he rose through the ranks, he remained focused on the needs of wildlife. While in Washington, D.C., in the early 1970s, he helped draft the first policies for the Forest Service in implementing the Endangered

Species Act. He was also on the front lines of the development of riparian management rules within the Forest Service. This didn't make him very popular. "No one valued riparian health back in the 1970s," he recalled. "One forest supervisor told me to my face to get lost. He said there were no riparian areas on his forest. It was all about timber and cows."

The unofficial attitude toward wildlife wasn't much better. Few biologists were employed by the agency, and the ones there were got caught up in intense turf battles with state wildlife agencies and the U.S. Fish & Wildlife Service. "The old thinking was: get the range right [i.e., grazing management], then the wildlife will be ok," he said. "In the old days 'wildlife' meant deer and elk, not much else."

Things began to change, however, mostly as a result of intense pressure from an environmental movement that was flush with victory at the time. Standards for what constituted healthy ecosystems rose, especially for riparian areas, but so did conflict among various interested parties, which raised the stress level among federal employees. In the late 1980s, as the chief wildlife biologist for the Forest Service in the Southwest Region, Bill got caught in the crossfire between activists on both sides of the endangered species issue. He tried for a while to walk a middle ground, but soon exhausted himself from the constant friction.

By 1990, Bill was done arguing. He retired and tried to relax. But a personal tragedy and an enduring desire to make things better pushed Bill into his new career a few years later. Upon completing a series of classes with hydrologist and restoration pioneer Dave Rosgen, whom Bill credits with organizing his own ideas, Bill was asked by Tom Morris of the Navajo Environmental Protection Agency to take a look at a serious erosion problem that endangered the western edge of Hubbell Trading Post, a National Park Service Historic Site near Ganado, Arizona. The straightened, rapidly eroding creek, which carried tons of sediment from poorly managed lands upstream from the Trading Post, was threatening the historic building. The park and the Navajo EPA were willing to give Bill's unconventional ideas a try, and this became the first project where Bill tried "induced meandering" on a practical scale. The creek responded quickly, and by 2007 had sufficiently "remeandered" itself for the Trading Post's structures to be considered safe from further erosion.

After his start at Hubbell, Bill fiddled with all of his ideas for riparian restoration over the next few years, only giving them a proper working out while on a consulting job in Mexico in 1996, which he liked a lot because there was no paperwork. Later, back in the states, Bill's ideas were greeted with a mixture of skepticism and outright resistance. Induced meandering in particular was a hard sell to regulating agencies, such as the Army Corps of Engineers, which controls the permitting process for stream restoration across the nation. That's because Bill's strategy was based on the principle that creating more erosion—by inducing a meander—was necessary to stop erosion. Additionally, Bill's emphasis on "sticks and rocks" as a way to fix creeks, instead of backhoes and cement, was a difficult novelty for many to accept.

Over time, however, as Bill's work proved itself where it mattered, on the back forty, the skepticism faded away.

Today, nearly eighteen years after his retirement from the Forest Service, Bill has never been busier. He is, in fact, booked. He has worked across the Southwest and beyond, and many of his ideas and techniques have been picked up by a new generation of restoration specialists. Of all the indicators employed to monitor the success of his work, this may be the most telling.

For all its practical applications, Bill's work also represents an important "splitting of the difference" between the "humans-know-best" paradigm that had dominated federal natural resource management and the "leave-it-alone" paradigm that dominated environmentalism. As it turns out, the middle ground that Bill sought years ago has been the most fertile.

So, the next time you see a trail beginning to dig into the soil, consider one of Bill's ideas: place an obstacle in the trail and make the cows, or people, meander around it.

Only, don't think of it as an obstacle—consider it an opportunity.

Chapter Ten

THE GIFT

Comanche Creek *Carson National Forest, New Mexico*

Taped to my computer is a postcard I found in a local coffee store. It depicts an ill-looking planet Earth, with its tongue hanging out, imprinted with the message: "The world could be in better shape." Surrounding this image are words: renew, heal, reaffirm, nurture, rekindle, revitalize, repair, revive, mend, soothe, rebuild, fix, regenerate, reinvigorate.

I've thought a lot about those words over the years, especially as The Quivira Coalition embarked on a series of riparian restoration projects under the direction of Bill Zeedyk beginning in 2001. By then, they were familiar words to me; I had heard them employed by Jim Winder, Kirk Gadzia, Sid Goodloe, Dan Dagget, and many others to describe their work and the work of fellow healers. They are words of advancement and action—positive, progressive, healing action. By contrast, much of the vocabulary I learned as an environmental activist focused on defense or safekeeping: save, preserve, roll back, stop, protect, prohibit, enforce. This vocabulary is still needed, but I've come to believe that it is the language of *healing* that gives people more meaningful direction and hope. I believe that because I've seen it in action over and over.

People respond to restoration work because it involves us in a "giving" rather than a "taking"—a giving back to nature, an honoring, while we necessarily continue to take nature's bounty. We can't stop using nature—we need its air, its water, its food, its animals, its minerals, its beauty, and its inspiration for our well-being. We must take, but *how* we take, as well as what we do with what we take and what we leave behind, lies at the root of many of our environmental troubles. Too often, we take too much. As we take, however, we can also give, and not just for the gesture's sake. Giving is becoming a requirement. The world not only *could be* in better shape—it *must be,* and soon, according to many experts and elders. The survival of the earth's living things (including us) requires that we renew, heal, reaffirm, nurture, rekindle, revitalize, repair, revive, mend, soothe, rebuild, fix, regenerate, and reinvigorate the planet's natural heritage.

But there is another reason why I like these words, something beyond the practical and the doctoring. They are words of redemption. We have taken so much from the natural world that simply cannot be put back again, ever. It's not just the passenger pigeon and the dodo bird either. It's something bigger, something to do with the way we treat each other, the damage we do to relationships with one another and with nature, as well as the trouble we cause natural processes. We sin without seeking redemption—and not just from the natural world. We rarely look for redemption at all levels of our daily lives, mostly because we don't have to. We live in an age and a society that has almost completely buffered us from the consequences of our actions. We take, we eat, we drink, we travel, and we consume—we sin—without retribution. Furthermore, a wide variety of cultural agents—including the TV, the grocery store, the automobile, the city—tell me I don't need to worry about giving back. Their message is clear: keep taking. All is well.

All is not well, of course. But I knew that going in as an environmental activist. What I didn't know, and what I learned over time as I traveled and met healers of all stripes, is that we *can* make things better, not by shielding a special place from all this taking going on but by *giving,* and in so doing try, even with small gestures, to redeem ourselves. In other words, the restoration of health—to creeks, grasslands, ourselves—is a moral exercise. I'm not sure that Bill Zeedyk, or any number of ranchers I know, looks at it quite in those terms, but I do.

So does William Jordan. In his book *Sunflower Forest: Ecological Restoration and the New Communion with Nature,* the former director of education at the University of Wisconsin's famous arboretum—where Aldo Leopold inaugurated the first formal restoration work in the history of the nation—Jordan argued that restoration is a form of "gifting" back to nature, an exchange with a moral purpose. "Everything we have, we take from nature," he wrote, "sometimes by persuasion or collaboration, sometimes by outright theft. Either way, the debt we incur is, or at least ought to be, a constant concern. For many, restoration is an attractive idea because it offers a way of repaying this debt."

Jordan considers restoration to be a gift both in the restored ecosystem itself and in the greater understanding and self-awareness that restoration creates among its practitioners. It is a redeeming gift, a gift of reciprocity; we give so that nature may give back. Under our old way of thinking, however, reciprocity was unnecessary. Nature did all the giving. "We accept its gifts of food, materials, place, and beauty," Jordan wrote, "but never offer back the clinching gift that would establish a basis for solidarity . . . and because we never risk the offering of a gift, we have no need for sacrifice."

He suggested that the old environmental movement shares much of the blame for the lack of gifting we do as a society. The movement has swung unproductively, he said, between insisting on a general withdrawal from society as a prerequisite for a meaningful relationship with nature, expressed in a philosophy of preservationism, and a utilitarian, dispassionate conservationism, expressed in timber yields and pounds-per-acre of grass productivity. As a consequence, the environmental movement failed to inhabit a middle ground where people could transcend the differences between these two approaches, and thus create a community. "This, I believe," wrote Jordan, "is why environmentalists have argued for community but have made so little progress in the task of strengthening the human community, much less expanding it to include other species."

Restoration can lead us to this middle ground, Jordan said, because it is not just about giving back to nature; it includes the gifts we give each other as we do the work—the honoring of different beliefs and ideas—which creates a cycle of giving and giving back that is as old, well, as the human species. Jordan cited the work of the great French anthropologist Marcel Mauss, who observed that the reciprocity inherent in gifting—I give so that you may give back—is one of the core elements of being human. It is the foundation of lasting relationships. Reciprocal gifting is different from charity or commerce and thus creates a deeper relationship that Mauss called "solidarity." Gifts are exchanged between collectives, not individuals, and the exchange is public, not private. And it does not "settle" accounts. It goes on and on—or it should, anyway.

But what sort of gift are we talking about? What do we mean by "restoration" anyway? Is restoring Glen Canyon by tearing down its famous dam, as author and provocateur Edward Abbey insisted, a reciprocal gift? Or is it an attempt at settling accounts? Would tearing down the dam in the Hetch Hetchy Valley in Yosemite National Park, whose construction in the early days of the twentieth century broke John Muir's heart, be an act of solidarity with nature? How do we "fix" a suburb of Phoenix? Or an overgrown forest? An overgrazed meadow? Are these acts of redemption?

Maybe. It depends on how we go about it, said Jordan. That's because it has to be about creating a relationship, not simply implementing an action, and it has to be about learning, stepping back, listening, waiting. "At its best, restoration can become an intimate dialogue with an ecosystem. . . . Restoration is work and it can also be play, a way of communicating with other species and with the landscape, a mode of discovery and a means of self-transformation—a way of both discovering the natural landscape and discovering ourselves in that landscape."

This is why Jordan considers restoration to be a form of agriculture—the restorationist prepares soil, controls weeds, grows plants, and harvests food, in a sense. But it is not agriculture as we commonly think of it. Although agriculture ordinarily involves bringing nature under control to a certain extent—simplifying an ecosystem to exploit it more effectively for some human end—restoration does just the opposite, Jordan said, "recomplicating the system in order to set it free, to turn it back into or over to itself, with a studied indifference to human interests."

In fact, restoration provides a way of redeeming agriculture (and thus ourselves) because it offers back to nature a gift in return for the gifts we receive from it. In doing so, Jordan concluded, restoration changes the relationship between nature and culture from negative to positive and thus opens the way to participation by large numbers of people.

By 2007, this made a lot of sense to me. But in summer 2000, when I received a phone call from Dick Neuman, who was then president of New Mexico Trout, a fly-fishing organization based in Albuquerque, redemption was the last thing on my mind. It was probably the last thing on Dick's mind as well.

He called because his group had labored for years, with only modest success, to restore Comanche Creek, a tributary of the Rio Costilla, located in the western half of the one-hundred-thousand-acre Valle Vidal unit of the Carson National Forest, in northern New Mexico. Dick and his fellow anglers desired to restore the creek to its former status as a prime cold-water stream for the Rio Grande cutthroat trout—one of only two native trout species in the state. As Dick explained, although there were plenty of "cuts" in Comanche, the population as a whole in the watershed was struggling for survival. Accepting his invitation for a tour, I quickly learned why the fish was in trouble.

The Valle Vidal had been hammered. Much of the West's plundered history, in fact, could be read into the condition of the property at the time of its transfer to the Forest Service in 1982 from the Pennzoil Corporation (which was the last in a series of corporate owners). Extensive overgrazing by cattle, scars from widespread logging and road building, and the "bleeding" effects of a historic gold mining district were visible all over this high, remote, and beautiful landscape.

As we toured the Valle Vidal that summer, Dick said after the transfer, the Forest Service, the grazing permittees, and various wildlife organizations made an innovative effort to reverse this Old West legacy. The grazing association hired a herder to control livestock impacts, the fishing organizations planted willow and cottonwood poles along the creek to stabilize eroding stream banks, and the Forest Service constructed a giant elk exclosure to keep out hungry herbivores. Although these efforts were helpful, Dick said, they weren't enough; the cutthroat trout population in Comanche Creek and its tributaries continued to struggle. All the willow and cottonwood pole planting, for instance, had failed.

Dick called me because he was worried. Cutthroat trout, he explained, like cool, clear water with deep pools and overhanging brush, very little of which was in evidence on the creek the day of our tour. He had an additional concern. His group wasn't the only one worried about the plight of the Rio Grande cutthroat trout; a handful of environmental groups were threatening to sue the federal government to get the fish listed under the Endangered Species Act. Dick thought that a legal confrontation would be bad for the trout. He wanted to find another way instead.

Fortunately, the Surface Water Quality Bureau of the New Mexico Environment Department—and by extension the U.S. Environmental Protection Agency (EPA)—was also worried about Comanche Creek. Excessive sediment movement (bad for a lot of reasons), the presence of aluminum (a toxic metal), and high water temperatures (bad for fish) had landed the stream on the state's list of impaired waterways, mandating action. The New Mexico Game and Fish Department was worried about the cutthroat trout too, specifically about all the nonnative brown and rainbow trout in Comanche Creek, outcompeting the natives.

The dialogue between these various interested parties, coordinated by The Quivira Coalition, resulted in an award from the EPA, under its 319 program (Clean Water Act), for a substantial multiyear grant to restore a portion of Comanche Creek to health. Partners included the U.S. Forest Service, the New Mexico Environment Department, New Mexico Game and Fish, Trout Unlimited, New Mexico Trout, the Rocky Mountain Youth Corps, the Philmont Boy Scouts, the Taos Soil and Water Conservation District, and consultants Bill Zeedyk, Steve Carson, and Kirk Gadzia.

As part of the 319 application, the partners, now called the Comanche Creek Working Group, agreed to the following process:

1. Conduct an assessment to identify specific impairments.
2. Conduct baseline monitoring and mapping.

3. Identify and implement best management practices.

4. Conduct an educational program.

"The ultimate goal of this project is to improve the condition of the Rio Costilla watershed to meet current water quality standards and to restore normal hydrologic function to the Rio Costilla and its tributaries," we wrote in the grant. "Completely achieving this goal will likely take decades. Over the next three years, however, we hope to establish the technical and organizational foundation for achieving this goal and to begin some on-the-ground restoration at Comanche Creek to maximize habitat for the Rio Grande Cutthroat Trout."

That was the official line—unofficially, we wanted to find a bit of redemption for the Valle Vidal.

In summer 2002, members of the working group conducted an assessment of the watershed. Their findings confirmed what long-time observers had suspected. The watershed suffered from three broad ills:

- The legacy of historical overuse was evident in raw stream banks and overall poor hydrological function, which contributed to high sediment loads in the creek.
- Poorly designed and maintained roads, including the main road, also contributed significantly to sediment transport.
- Overgrazing by cattle and elk in spots were prohibiting the growth of shade-creating woody plants, such as willows and cottonwoods.

After baseline monitoring and mapping were completed, the working group embarked on a three-pronged strategy to address these problems.

RAW STREAMBANKS

Under Bill Zeedyk's guidance, we proposed to construct a large number of erosion control structures within the watershed, including wooden vanes and baffles in the creek itself, one-rock dams in the uplands, and worm ditches, rock baffles, and headcut control features in wet meadows. The purpose of these structures was to speed up natural recovery processes. Scouring by erosion from historic overgrazing and logging resulted in the creek cutting down below its natural floodplain. Over time, the creek began to heal itself by creating a new floodplain—"remeandering" to dissipate energy and drop sediment—but there were plenty of old "wounds" that had not healed. The goal of the restoration work was to "goose" the healing process along gently.

BAD ROADS

Bill led an inventory of the roads and prioritized which needed attention first. He paid particular attention to the placement of culverts; a poorly placed culvert can quickly create a headcut and cause erosion. Bill pointed out how much water was being trapped in roadside ditches, thus starving downslope plants. This water also gathered a great deal of sediment as it picked up speed in long runs downhill. He proposed that many of these roads receive "rolling dips" so that water would be allowed to flow again in their "microwatersheds."

HUNGRY ANIMALS

Although the cattle were largely controlled by a range rider, our monitoring showed that cattle-caused "hot spots"—usually areas that had been excessively trampled—still could be found in the riparian area. Elk had grazed the area too. And both animals are very fond of young willows and cottonwood trees. In the mid-1990s, the Forest Service experimented with a novel idea: create house-sized "mini-exclosures" around existing native willow clumps to protect them from grazing animals. This was in contrast to the big elk exclosure built in the 1980s on the creek, which proved difficult to maintain.

The mini-exclosures were deemed a success. So the working group, with the energetic assistance of the Rocky Mountain Youth Corps, employing "at-risk" youth from the Taos area, decided to build many more elk exclosures. The goal was to protect the willows so they could grow and shade the water, thus reducing overall stream temperatures—a critical requirement for the fish.

With the grant in hand, and our partners in place, we set to work.

Five years later, in July 2007, I decided to take a walk. I was back in the Valle Vidal as part of a restoration workshop on the middle reach of

Comanche Creek, the latest in a long series of productive outings. This time we had volunteer help from the Sierra Club, New Mexico Trout, and numerous others—more than sixty people in all. The goal for the weekend was to construct nearly two dozen vanes—wooden posts that are placed in the creek at a wedge-shaped angle to the bank so that water is deflected away from the eroded edge of the creek. They're quick to build and fun to install, because it involves getting muddy. Before long, everyone was busy hauling, hammering, or digging. Laughter and high spirits filled the meadow.

After lunch, Bill Zeedyk led a tour of our restoration work in nearby Holman Meadow, which was suffering from a sequence of ugly head-cuts, so everyone eagerly trooped off after him. Everyone except me—I stayed behind. I had heard Bill's talk many times, but that wasn't why I stayed. I wanted to walk down the creek, all the way to Comanche Point, five or six miles away, and see for myself what we had accomplished. Five years and one grant extension later, I knew that we had accomplished a lot. I knew the numbers, anyway: more than 130 in-stream structures constructed, repairing a total of more than 35,000 feet of channel length; fifty elk exclosures built, enclosing nearly 200,000 square feet of stream bank; and more than 100 upland sites treated, including portions of many abandoned roads. I also knew from our annual monitoring that conditions in the creek and the uplands were improving, dramatically in some places. Banks protected by vanes were revegetating; willows and even an occasional volunteer cottonwood were growing steadily in the exclosures; culverts were replaced; road scars were healing; and the creek itself was narrowing and deepening in places—which is a very good sign.

I also knew that the U.S. Forest Service and the New Mexico Game and Fish Department had finally constructed a fish barrier on Comanche Creek, a prerequisite to the eventual restocking of cutthroat trout after a purging of the nonnative species by piscicide (sometimes healing means hurting, alas). In other words, Comanche Creek was well on its way to becoming a healthy cold-water fishery again for the cutthroat. Dick's dream was becoming a reality, it seemed. The numbers were good, but what did it *look* like?

I started walking.

The first thing I saw were eroded streambanks. In many spots, usu-ally on outside bends of the stream, three-to-five foot tall raw wounds of exposed soil punctuated an otherwise picture-perfect vision of a verdant meadow. Peering down I could see places where large clumps of soil, topknotted with tufts of grass, had collapsed into the water as if weary from the effort to remain upright. Peering closer in places, I saw slashes of blue on the banks and sandbars—spray-painted directions by Bill. A vane would be created here, a baffle there. The slumping would eventu-ally stop. The weariness would be replaced, eventually, by vigor.

I walked on. At a culvert below the road, I stopped and pulled out my camera. Five years ago I stood on this very spot, snapping the same one-two sequence of photographs that I was doing again. I could tell already that much had changed. What had once been a nasty, gaping wound below the culvert, which shot a large plume of soil into the creek below where I stood (very bad for fish) was now grassed-over, as was the plume. Downstream a short distance I could see a wooden vane—seven or eight wooden posts placed vertically in the creek—protruding upstream from the edge of a large, raw streambank. The "pillow" effect created by water pooling behind the vane had already begun its work: the creek was turn-ing, slowly but steadily, away from the bleeding bank. The wounds were closing. I took photos.

I kept walking. One criticism frequently leveled at restoration by environmentalists is that because we "don't know what we're doing" we run the risk of creating a cure that could be worse than the disease. Better to do nothing at all—and let nature do the mending. This is a fair shot—but I think it is a decade late. A medical analogy is useful again: a century or more ago medicine was still mostly guesswork, and some-times the guesses were wrong, with deleterious consequences for the patient. Today, of course, thanks to huge, amazing strides in research and technology, few would argue that the cures provided by modern medicine are worse than the disease. I think restoration is approach-ing the same threshold, thanks to pioneers like Bill Zeedyk, Dave Rosgen, and many others. Are we still guessing with our restoration activities? Well, not really. The science of hydrology, for example, is rather sophisticated now, as is our knowledge of soils, and plants. The proof is in the pudding—as I could see all along Comanche Creek: the patient was recovering. We don't know everything, of course, but we know enough to get going.

Certitude is a funny thing, I thought as I walked. I recalled a conversation I had with a reporter two years earlier. He called to say that a prominent environmentalist in Arizona, one of the litigants in the drive to get the cutthroat trout listed under the Endangered Species Act, had told him we were wasting our time on Comanche Creek. "Sticks and rocks will never save the cutthroat trout," he told the reporter emphatically. "What did he suggest instead?" I asked. Get the cows out, the reporter said, and then reintroduce the Mexican wolf. "Really?" I replied. "I didn't know wolves built erosion control structures."

I kept following the creek. Crossing the road at Little Comanche Creek, I recalled images from the spring flood of 2005, which had ripped apart many of the elk exclosures we had constructed on this lower stretch of Comanche. As it turns out, there had been a miscommunication between Bill and the construction crews—the wire fences of the square exclosures, most of which straddled the creek, were built too low to the water and were not braced properly. The floodwaters pressed against some of the fences until they collapsed in big heaps, pulling the posts at the corners toward the center of the exclosure like big matchsticks. It was quite a mess, and not a little discouraging. But Bill and Steve and everyone involved said, "Let's get back to work." And by the end of the summer every structure had been repaired. "Welcome to adaptive management!" we joked. In reality, it was the power of redemption at work. We weren't about to let our gift be taken away.

There had been other setbacks. The grazing association that used the Valle Vidal during the growing season had proved, somewhat to our surprise, to be uncooperative. Although they employed a herder, who was doing a reasonably good job of keeping the cows out of the riparian area along Comanche, he was not watching them closely enough in some of the side drainages. Our monitoring found cattle-caused "hot spots" that were creating potential new sources of erosion. It wasn't *bad*, but it wasn't good either. Moreover, the cattle part of the equation could be cured relatively easily, we thought. But when I called the head of the grazing association to suggest a meeting, he lost his temper. He accused me of not knowing "which end of a cow got up first" and hung up (it's the back end).

I kept going. The land along the creek looked great. The willows were tall and healthy behind their fences. The sky was a brilliant blue, the sun high and bright. The day was warm with the fullness of summer. I stopped at each structure, marveling at the creek's innate ability to renew, heal, reaffirm, nurture, rekindle, revitalize, repair, revive, mend, soothe, rebuild, fix, regenerate, and reinvigorate, if given a chance. I felt reinvigorated too. Even though we had only scratched the surface here, so to speak—the scale and quantity of work to do in this one watershed alone could probably occupy a lifetime—it felt good to see so much restoration taking place. But I knew the taking would never cease—not as long as there are humans on the planet. The issue, it seemed, was how to balance the giving with the taking.

But at that moment, I didn't want to think about that. I wanted simply to revel in the signs of renewal. The Valle Vidal, with its legacy of hard use hopefully finished forever, is now writing a new, more hopeful chapter—a chapter employing the language of land health and healing. Moreover, with its new history of restoration, it has the chance to become a leading landscape in the new movement to reconnect people to nature meaningfully and adaptively. We can't fully atone for the sins inflicted on this beautiful place, but we can heal old wounds, and in the process heal ourselves. The Valle Vidal.

The Valley of Life.

PART THREE

The Big Picture

Chapter Eleven

ACROSS
BOUNDARIES

Laramie Foothills *North of Fort Collins, Colorado*

One glance at the overflowing bookshelves and coffee tables in Rick and Heather Knight's home tells the visitor that they are up to something big.

The books and the stacks of articles range from scholarly histories and professional papers on conservation biology to environmental journalism and humorous cowboy stories, but the theme is the same: the American West, big. In fact, the scale of the Knights' professional and personal focus is neatly summarized in the title of one of Rick Knight's own books: *Stewardship Across Boundaries*. And as the title suggests, much of this focus could be called unorthodox.

Still, it is not a coincidence that one of their favorite authors is the late Wallace Stegner, award-winning novelist, essayist, historian, and iconic dean of the region. Like Stegner, the Knights are passionately devoted to the West, to its history, its wildlife, and its wide open spaces, and also like this famous author, they are determined to help the region heal its wounds and become healthy again.

It hasn't been easy.

"We live in a take, take, take world," said Rick. "Even recreation is a form of taking, though most people don't think of it that way. What we need to do instead is give, and that means becoming involved in a place-based conservation effort."

This is exactly what the Knights have done. In their case, the place is the mix of private and public land that stretches twenty-five miles north from Fort Collins, Colorado, to the Wyoming border and approximately thirty miles west from Interstate 25 into the mountains—the last significant block of relatively unfragmented land left along Colorado's Front Range.

Considering what has happened to the land from Fort Collins to Colorado Springs, one hundred and fifty miles or so to the south, and then beyond—gone to subdivisions at the rate of an acre per hour—the Knights believe the only hope for their community is through a collaborative effort that unites ranchers, farmers, conservationists, scientists, and local officials.

"There is no other way," said Heather. "Either we link arms to protect this precious landscape, or we're going to lose it forever."

United in their goal, each has chosen different paths of action. Heather works for the Nature Conservancy (TNC) as director of the Laramie Foothills—Mountains to Plains Project, which aims to conserve the private land in the "danger zone" through a combination of conservation easements, tax incentives, land purchases, and stewardship initiatives.

"Heather is the keystone in an incredible community effort," said her husband. "No one I know works harder, or has had more success."

Rick Knight is no slouch himself. A professor of wildlife biology since the late 1980s at nearby Colorado State University (CSU), Rick has eight books and more than 110 peer-reviewed articles under his belt, most of them focused on land use and conservation. He has sat on numerous boards, including that of the Society for Conservation Biology, and remains involved in active research, not to mention his full-time teaching duties. And in 2007, the university awarded him its highest honor for teaching excellence.

Neither Rick nor Heather is native to the West, but that doesn't make them any less "western" than many other worried residents, including Wallace Stegner, who was born in Iowa. In fact, like Stegner, their experience in "becoming western" has provided rich material for their careers as well as their concerns.

Rick was "born on wheels," as he put it. His father, a navy scientist and world authority on mosquito-borne diseases, moved the family from Maryland, where Rick was born, to Florida, then to Egypt and beyond, meeting career demands.

"More Marines died in World War II from infectious diseases than were killed by the Japanese," said Rick. "That's why the navy kept dad so busy."

In addition to cultivating a desire for lifelong learning (there are five Ph.D.'s in Rick's family), his father's agrarian roots also had a profound impact on the future biologist. "I grew up with a healthy respect for rural people," said Rick, "which only grew stronger as I got older."

After a tour in Vietnam as a Marine himself (a platoon commander in the Demilitarized Zone), Rick returned to the states to pick up his education. A master's degree in wildlife biology at the University of Washington was followed shortly by a Ph.D. from the University of Wisconsin at Madison, the stomping ground of another of Rick's heroes, Aldo Leopold. An offer of a faculty position at CSU, and a move to Fort Collins, came next.

Through the years of traveling, studying, writing, and teaching, a common thread to Rick's work emerged: knowledge, to be relevant, needs to be place-based and applicable to the real world. This has frequently set him at odds with his colleagues, especially in academia, who often are reluctant to bridge theory and practice.

"There is no reason for our teaching or our research to be esoteric," said Rick, "it needs to be relevant. And considering the world's conservation problems, what could be more important than teaching, service, and research that benefits both human and natural communities?" About his adopted home, he feels a special urgency. "Given the rate of loss of private ranch and forest lands in the West," he said, "what we accomplish in the next ten years will be critical in how the West looks and for the long-term survival of much of the region's natural heritage."

Heather Knight traveled more miles, and in less time, than her husband did to get to Colorado, which might explain why, despite more than fifteen years of steady work on land management issues, she is not ready to call herself a native.

Another reason might be because a remnant of her Australian accent identifies her as "exotic" still.

Although Heather grew up in Sydney, she had family farming roots, which is why when she attended college (the first in her working class family to do so) she gravitated toward the "applied" side of her ecological studies. Desiring to be outside as much as possible, and needing a paycheck, she worked every summer in the field doing biological research, which also allowed her to exercise another passion—for long-distance running.

After a postgraduation stint in the recreation industry at Alice Springs, Heather headed to Brisbane, where she earned a graduate degree in science education. A teaching job was followed by more graduate research, this time focused on the impact of tourist helicopter overflights on the marine wildlife of the Great Barrier Reef, which the data demonstrated to be substantial.

Birds were another passion. In 1990, she accepted an invitation to deliver a paper at the International Ornithological Congress in New Zealand, where she met an exuberant American wildlife professor from Colorado, also there to report on recent research. Five days after making introductions, Rick Knight proposed.

"Fort Collins was a total shock," she recalled. "All the snow, and no humidity. The first time I went for a run I got a nosebleed. And it felt like I had a heart attack, because of the elevation."

She applied herself to the study of her new home. She volunteered for TNC and was quickly promoted to a full-time position at the conservancy's nearby Phantom Canyon Preserve.

At the time, TNC was going through significant changes, including a transition to a community-based model of conservation, and recognition that effective ecological protection meant working at the landscape scale. Both of these changes resulted in a decision to hire staff from within a community as well as to begin a collaborative process that could engage a wide variety of landowners and interested parties in a landscape.

"We found out that if conservation were only about money, it would be easy," she recalled. "But it's mainly about relationships and working across boundaries with neighbors, especially if you want the results to be longer lasting."

Her work took her into the homes of local ranchers and farmers, where her diplomatic skill helped her win their trust. She listened, too, to what community members said was needed. For example, she turned her prodigious energy to the creation of a local weed cooperative, needed to combat one of the West's most intractable challenges: the cancerous spread of invasive, noxious weeds.

"Weeds are the third biggest threat to biodiversity in the West," said Heather, "so it's got to be taken seriously. And the thing about weeds is that it requires cooperation across fence lines. A single landowner can't do it alone."

Through countless meetings, and endless cups of coffee, Heather marshaled the weed cooperative into existence. To her surprise, her accent was not a liability. "I'm an alien too," she said with her easy smile, "but I'm not invasive or noxious, I hope."

In late 2004, all the trust-building and hard work paid off as the community celebrated a major success. It involved the four-hundred-thousand-acre Laramie Foothills, a biologically rich transition zone that stretches from shortgrass prairie in the east to alpine mountains on the west. Notably, more than 70 percent of this landscape is privately owned, predominantly by a few remaining ranch families, and nearly all of it is threatened from contagious subdivisions to the south. In the mid-1990s, TNC decided that this twenty-two-by-thirty-mile block of land was a top conservation priority.

One reason was wildlife. As Rick often points out, most western mountain ranges run north-south, which means most migrations are east-west as wildlife move from summer to winter pastures and back again. When ranches constitute the critical "in between" lands, as they do in the Laramie Foothills, the migrations are unobstructed. But when subdivisions spring up, wildlife migration routes are severed, often to deleterious effect. "Up and down the Front Range, the land was being broken into smaller and smaller pieces," said Heather. "We all understood that the Laramie Foothills was our last, best chance to protect critical wildlife migration corridors."

One of Heather's main tasks was to reassure private landowners that the word "work" would remain a part of "working landscapes"— something that was met with skepticism initially.

"Fortunately, being a foreigner I genuinely wanted and needed to learn as much as possible," Heather commented, "so I asked if I could come to their place and learn. I was also willing to admit mistakes. Both of these created a different kind of conversation with the landowners here."

Over time, she helped place fifteen thousand acres under conservation easements, both through TNC and a local land trust, thus protecting the land from development in perpetuity. She also worked hard to get local governments more actively involved, assisted by new tax credits that allowed for the purchase of open space.

"There's nothing more satisfying than seeing a diverse community come together for a shared larger goal," said Heather. "Now we have families involved who absolutely stayed away in the beginning. It takes time, but the payoff in terms of conservation is bigger."

Then the final piece fell into place. An $11 million proposal by Heather and friends was approved by Great Outdoors Colorado, a state agency that uses lottery money to protect open space. Fifty-five thousand acres of private land will be protected, 65 percent by easement and the rest by purchase. Money for additional protection will be leveraged with TNC, county, and city money, protecting another twenty thousand acres of private land from development.

By 2007, the news was even better. "With the work of a diverse and highly committed partnership," reported Heather, "we now have 100,000 acres of private lands conservation completed and connected to 110,000 acres of public lands stretching across our watershed, committing forever 50 percent of this landscape to conservation and compatible land uses, such as ranching, literally stretching from the mountains to the plains."

But the work is not done. Protection from development is just the first, big step, as Heather noted. Demonstration projects need to be completed, capacity building needs to be pushed forward, grazing plans need to be developed and implemented, water resources need to be secured, reliable flows in the North Fork of the Poudre River need to be restored, prescribed fires need to be designed and lit, education and outreach programs need to be funded, partnerships need to be expanded, new tools need to be added to the stewardship toolbox, and, of course, weeds need to be pulled.

Still, the Laramie Foothills project represents a significant achievement, as well as a precedent. "This is the first project in the West where TNC hit at least 85 percent of its goals," said Heather. "A skepticism about community-based collaboration I hear all the time centers on how long it takes," she said. "But as I think we've proved, it's the only way to get large-scale conservation done. And as my mother liked to say, patience is a virtue."

But perhaps most important to Heather was the trust and cooperation she earned from the landowning families. "It was incredibly hard," she said. "All the families were under incredible pressure, including pressure from internal conflict. But it was such a joy to work with them all these years."

Protecting large landscapes can seem like a quixotic adventure. But this is precisely what appeals to Rick Knight, who has a bit of the *Man from La Mancha* in him. That's because, like the famous Don, this professor of wildlife biology tends to charge at many of the West's most sturdy windmills.

Take the issue of recreation on public land, for instance. Most environmentalists and many federal managers tout recreation as the "highest and best use" of our public lands. Recreation is also promoted as the principal economic hope for many rural communities, especially as the number of day-visits to our national forests and parks continues to rise, to more than 1 *billion* per year at last count.

"Thus, whether on foot, by horse, motorcycle, mountain bike, ski, or snowmobile," wrote Rick in an article, "people will increasingly enter our public lands to seek spiritual elevation, aesthetic enjoyment, the companionship of family and friends, exercise, or just to escape from the stress of our urbanized cementscapes."

"Isn't this ok?" he continued. "Hasn't this been the struggle that has defined the environmental movement for almost a century? Out with the damaging extractive uses of logging, mining, and livestock grazing and in with the more environmentally friendly and benign pursuits of outdoor recreation."

Now comes the charge at the windmill. To Rick, this shift may *not* be ok. "From where I stand, there appears to be a certain degree of duplicity in our discussions to substitute amenity uses for commodity uses."

That's because people are not asking two tough questions: first, whether or not recreation is ecologically benign; and second, whether we can better manage recreation than we did logging and grazing.

On the second question, Rick is not sanguine. Trends in tourism and rural economic development, recreational equipment and off-road vehicle sales, and federal land policy goals all point in one direction: up. In fact, Rick noted, the number of people participating in wildlife-oriented recreational activities is projected to increase between 63 and 142 percent over the next fifty years. This news is worrisome to the biologist.

"Just as historically we overgrazed, overlogged, overmined, and over-dammed our public lands," he wrote, "today we are gathering together the forces that may overrecreate these lands in the future."

It is the answer to the first question, however, that most distresses him. Recent research, including his own and that of his students, suggests that the ecological effects of recreation are far from benign. Recreationists, noted Rick, create various deleterious effects: they modify vegetation and soil with their activity; their presence can cause the abandonment of preferred wildlife feeding sites; and the stress they cause wild animals can change reproduction rates, or even cause death.

Rick summed up the research by himself and his students this way: "Regretfully, this new American West with its robust tourism-dependent economy will result in an altered natural heritage. Rather than seeing more species that have figured prominently in our imagination of the West, we will see fewer."

In perhaps his most quixotic adventure to date, Rick has tackled the politically volatile question of biodiversity and land use in the West.

Under traditional thinking, biodiversity—the number and variety of species in a given area—is always highest in protected areas, such as national parks and wildlife refuges, especially when compared to working ranches or exurban (low density) housing developments. As a consequence of this thinking, the traditional response by conservation groups and others to threats to biodiversity has been to increase the size and number of acres in nonconsumptive use (parks and refuges—other than recreation).

However, the results of a study led by one of Rick's students, Jeremy Maestas, challenged this conventional line of thought. The study area was a blend of public and private land northwest of Fort Collins. The team sampled plant and animal communities by selecting points randomly across the landscape for study. Particular attention was paid to songbird populations.

They found densities of native plants and animals higher on ranches and protected areas than in exurban developments. Human-adapted bird species, for example, reached their highest densities in the exurban areas. Dog and cat populations were very high in these areas as well, while coyotes were rarely seen.

This is significant because dogs and cats, which the authors call "subsidized predators," have a major impact on wildlife.

"House cats have been implicated in the decline and extinction of scrub-breeding songbirds in two studies in California," the authors noted. Dogs are known to harass and kill wildlife. "Research has shown that they can extend the zone of human influence and contribute to the annual mortality of some species," they said.

More surprising was another conclusion. "More species of native plants were found on ranches than on the other two land uses," they wrote. "The dominant nonnative plant, cheatgrass, was more prevalent in terms of cover on the protected areas and exurban development than on ranches."

Ranches supported greater biodiversity than exurban development, they determined, for three reasons: ranches had fewer human-adapted wildlife species, greater numbers of native wildlife species, and more native plant species than exurban developments.

The coauthors attribute some of these findings to the higher elevations and the less-productive soils. Additionally, many critical riparian areas, which were historically homesteaded, lie on the lands of private ranches. "Our results combined with this information suggest that we will not be able to sustain native biodiversity in the Mountain West by relying merely on protected areas," they concluded. "Future conservation efforts to protect this region's natural heritage will require closer attention being paid to the role of private lands."

Wallace Stegner once called the American West the "native home of hope." Few individuals in the region understand this optimism better than Rick and Heather Knight. At the same time, few understand the challenges better than they do.

"It's easy to despair," said Rick, "but being involved in community-based initiatives is the best antidote for pessimism I can think of." Rick finds solace and inspiration in his students. "I have been so blessed," he said. "Every one of my graduate students is working in conservation today. I love teaching. It's what keeps me going."

Rick led an effort to create a new, interdisciplinary department within CSU called Forest, Rangeland, and Watershed Stewardship. Its goal is to integrate what had been disparate disciplines and focus the course work on applied research and management. It is the first department of its kind in the nation.

"There was real resistance to the word "stewardship" within the university," Rick said. "Many academics still believe in the ivory tower. But the truth is, if we don't do something in the real world, we're going to lose most of what we love."

Following his own advice, Rick joined the board of the Colorado Cattlemen's Agricultural Land Trust, an industry-led group dedicated to preserving ranching through the purchase of conservation easements and other emerging tools. He also serves on the science advisory team for the Malpai Borderlands Group, a collaborative nonprofit organization in southeastern Arizona.

"At the end of the day, all of us who live in the West should pause and consider how fortunate we are," said Rick. "It reminds me of E. B. White's quote, 'I rise every morning torn between a desire to save or savor the world. This makes it hard to plan the day.'"

"What I've learned," said Heather, "is that home is not a place where people live, it's a place where people care—about the land, each other, and a shared future."

Chapter Twelve

USEFUL SCIENCE

Jornada Experimental Range *Las Cruces, New Mexico*

The first time I heard Dr. Kris Havstad give a presentation on desert ecology, he began his talk in the back of the room—literally.

His point was figurative as well as literal—that for too long scientists were most comfortable in the back of the room, listening attentively, but politely disengaged from the controversies surrounding natural resource management in the West. The reasons for this detachment, he said, included a concern about incomplete knowledge, a fear of getting dragged into politics, an aversion to conflict, and even a certain shyness.

As Kris enumerated these reasons, he walked slowly to the front of the room. This was also meant to illustrate a point—that it was time for scientists to be out front, to be engaged, to be useful. In Kris's opinion the main role of scientists, especially government scientists, such as those who work under his supervision at the U.S. Department of Agriculture's Jornada Experimental Range (JER), is to deliver their knowledge in ways that have impact in the real world.

"If we're not relevant," Kris said recently, "then the public should bag us."

It is a typical statement from a man who has gained a well-earned reputation as an articulate spokesperson both for the role of science in society and for explaining how arid ecosystems work. An opportunity to combine the two was one of the reasons Kris left a professorship at Montana State University and took the job as the boss of the JER in 1994.

"The tradition within the research community has been to publish," Kris continued. "Today, the goal is to deliver and have impact."

When Kris began walking the talk, which included becoming a founding board member of The Quivira Coalition, the idea of relevant science was considered somewhat radical, especially the holistic (meaning "whole systems") approach taken by the Jornada scientists, which contrasts with the "disciplines" approach, in which specialists pretty much stayed within their chosen fields.

Moreover, the tenor of the times was not exactly ripe for thoughtful analysis. The political climate was confrontational, the struggle between ranchers and environmentalists had reached a crescendo, and antifederalist feelings in rural counties ran high. It's no wonder scientists felt safe in the back of the room.

It took some courage, therefore, for the JER researchers and others to step forward. Perhaps the best way to explain it is to repeat a phrase I heard Kris use often when opening a meeting with ranchers and other private citizens at the time: "We're from the government, and we're here to help."

Summarizing the wealth of rangeland science today is as difficult as summarizing the desert itself. The more we learn, it seems, the more we realize how little we know. And if the goal of achieving ecological understanding isn't enough, add to it the job of communicating this knowledge to growing, and diversifying, lay audiences and you have a recipe for a Tall Order.

Adding to the difficulty is the changing nature of rangeland research itself. Until recently useful science at the JER meant research that supported agricultural aims, specifically the goal of raising cattle in an arid environment. In fact, the principal reason the 190,000-acre Jornada Ranch, located north of Las Cruces, became government property in 1911 was so scientists could study the damage done by overgrazing in the desert and consider mitigation strategies.

There was a lot to consider. The boom years of the livestock industry (1880–1920), characterized by large numbers of animals and few controls, had decimated the range across the Southwest. Throw in periodic drought, and you had a serious problem.

Widespread alarm at the time over soil erosion, loss of vegetative cover, and other grazing-related maladies stirred the U.S. government to take remediative action, including the creation of the Soil Conservation Service, to assist private landowners. It also set researchers, such as those at the new Jornada station, to work on trying to come up with a better way to raise livestock in the desert.

In the beginning the young discipline of rangeland science was directed to maintain focus on livestock, especially on the issue of increasing forage production. But over time, researchers began to see that it was a lot more complicated than simply trying to figure out how to feed livestock in the desert.

"Rangelands involve thousands of variables and millions of interactions among those variables," said Kris. "They don't 'behave' in predictable ways; they defy easy, quick, simplistic solutions or responses; and they challenge specific blueprints for their management. That's the conundrum. We need years of scientific study, but the land manager is expected to provide quick, correct, and practical answers."

One significant change brought about by the growth of ecological understanding in recent decades has been the broad shift within the science community from the idea of a "balance of nature"—in which natural forces are seen to be engaged in a constant effort to maintain equilibrium—to the idea of the "flux of nature"—in which disturbances, or "pulses of energy" such as fire, wind, and animal impact, are seen as natural occurrences and key to the maintenance of land health.

This shift in thinking has been accompanied at the JER by a parallel shift in the definition of useful science from something used primarily by industry to something that a whole community could use. "It's not about creating more forage for cattle anymore," said Kris. "Our mission is to improve our knowledge of ecosystem processes as a basis for management and remediation of desert rangelands."

Believing that a basic understanding of ecological processes is a prerequisite for prudent decisions by land managers, the researchers at the JER started over, in a sense, at the level of soil, grass, and water.

"What we've learned since the 1950s," said Kris, who like many scientists, tends to talk in lists, "is a better understanding of ecological principles, the mechanisms that drive ecological change, and the characteristics of ecological sites. That means today we should: one, understand the ecological processes in specific environments; two, know local conditions that modify those processes; three, monitor to evaluate responses [due to climate or management actions]; and four, adjust management."

One of the main tools that the JER scientists, in cooperation with many other researchers, have helped to develop is the Rangeland Health Indicator protocol (the system used to generate the Altar Valley map), a qualitative assessment that enables land managers to evaluate the effects of particular actions. They also helped to develop a quantitative monitoring system that precisely measures watershed function over time. In measuring basic ecological processes, these tools are designed to help answer real-life questions: what is the land capable of supporting, according to its soils and plant communities? What is the system lacking? What management strategy is appropriate?

It also means, according to Kris, finding "teachable moments" so that answers to these questions can be effectively communicated. One of the researchers on the front lines of finding these moments is Dr. Jeff Herrick. "It's all about better understanding the land and how our actions might affect it, and then communicating," said Jeff, a soil scientist at the JER and one of the principal authors of the rangeland health monitoring and assessment protocols. "You can develop the best tools, but if you can't communicate them then you aren't helping." Jeff does his part by leading numerous training sessions, mostly on the assessment system, for government land managers, conservationists, ranchers, and other landowners.

"The point of the training, and really of all our work," said Jeff, "is the assumption that what we value from land, such as livestock use or recreation or hunting, depends on three attributes: soil and site stability, hydrologic function, and biotic integrity. Without that healthy foundation you'll never reach true sustainability."

One of the messages that Kris voiced when he walked from the back of the room to the front, years ago, was that from a scientific perspective the "debate" over livestock grazing in the Southwest was largely over. He summarized this message in the November 1999 issue of The Quivira Coalition's newsletter:

"We now know that many of our rangelands have been overgrazed, that some areas remain in degraded states despite adequate rainfall, and that some rangelands shouldn't be grazed by livestock. Yet, we also know fairly clearly that livestock grazing of rangelands can be a sustainable practice for many sites, for many seasons, and for many years. Extensive experimentation has illustrated that grazing can be managed and the integrity of rangeland ecosystems—in terms of their ability to produce, capture and store nutrients and to conserve soil resources—can be maintained."

In other words, the ecological function of rangelands, if maintained, can support a societal value—livestock production. That was settled. The next step, he wrote, was to explore the ecological processes in more detail to provide the basis for their proper management.

What could not be done, however, was provide a "silver bullet" for successful livestock management in arid environments. To illustrate this point, Kris employed an analogy about the process of raising a child. Despite an impressive body of scientific knowledge about the medical, psychological, and physical elements to child-raising, Kris wrote, there does not exist a scientific "blueprint" for parents to follow—no single methodology that services all possible combinations of parents, children, and environments. Likewise, he continued, "although we have an impressive knowledge base for rangelands, there does not exist a single science-based blueprint for how we interact with our environment."

This is important for two reasons: first, it demonstrates that rangeland management, like parenting, will always be more art than science. In other words, science can inform, but not dictate, our decisions. Second, because these decisions will always be based primarily on societal values—culture, politics, economics—useful science means making as clear as possible the functions that undergird these values, much the way medicine and psychology are used to inform the raising of healthy children.

For example, the JER has begun recently to focus on another societal value: sustainability. Putting this value, which Kris defines as "the maintenance of ecological integrity over time," into practice is especially important, not only if we intend to continue to use arid landscapes into the future, but also for wider issues involved in the human/nature relationship. Therefore, the goal of sustainable use in the desert has major relevancy.

"Sustainable use can be defined as an appropriation of production, such as biomass used by grazing livestock, for instance," said Kris, "that allows for natural processes to replace appropriated materials."

Levels of use—all use—must be gauged by the natural limits of an ecosystem. That's the theory; the practice is more complicated. Useful science along these lines is difficult to achieve because although ecosystems exist at tremendous scales, both in time and space, we tend to view them, for obvious reasons, at human-friendly scales, which means the toolbox available to land managers is limited to only a few components of a landscape.

Such as plants. "There is much we can't control, but we can control plants," said Kris, "and we know a great deal about plants. Water too, especially the water cycle in arid environments. We can base our management actions on how we impact properties of these landscapes that are related to these key processes."

And that's where livestock grazing comes in—as a tool for the maintenance of key ecological processes. It is, in fact, one of the few tools land managers have, along with fire, rest, and certain forms of technology, to ensure proper functioning condition. We could choose to "leave well enough alone" too, letting nature take its course. But that's not a viable option in many places on the planet anymore.

Take desertification, for example—a sustainability concern for one-third of the globe. It is characterized by the unnaturally rapid loss of soil's protective plant cover, resulting in erosion by wind and water, which threatens the very processes that sustain life. If we are to reverse desertification—and many people think we need to—then we have to think about plants, as well as the proper use of land management tools that encourage their health, including well-managed livestock grazing.

What other roles might livestock themselves play in this emerging concern for sustainability? At the JER, this question is very much on the mind of Dr. Ed Frederickson, an animal scientist. For starters, he wants to know how beef cattle use landscapes and what their impacts are over time.

"It's an interesting question, in part, because there really aren't any specific answers," said Ed. "Like most ecological questions, every component of the system being studied is changing at a different rate. The situation is never the same at any two points in time. Animals are learning, their social interactions constantly shifting, and physiological needs are adjusting to varying internal and external conditions every moment. Likewise, their environment is even more dynamic."

So, why research a question to which there are no ready answers? Ed cites Simon Levin's book, *Fragile Dominion*, in which the first commandment of environmental management is to "reduce uncertainty." It's the same reason we watch the weather forecast on TV—for a prediction of the future. Increased certainty allows land managers of all stripes to improve their skills, which increases our ability to achieve sustainability. However, as Ed noted, we should be aware of Levin's second law of environmental management: "expect surprise."

The specific uncertainty Ed is trying to reduce is how livestock alter landscape soil nutrients, seed distribution, and the character of plant communities. This is important to know, he believes, because research has demonstrated that livestock hastened the conversion of 94 million acres of desert grassland in the United States into mesquite shrubland not simply because of overgrazing, but primarily because of mesquite seed dispersal through the digestive tract of livestock.

This type of research permits researchers at the JER to detect emergent ecological patterns or properties and predict future landscape directions given a range of potential scenarios.

That raises another question for Ed: what is the best cow for arid ecosystems? He suspects it might be Criollo cattle, a lighter animal than the English breeds that became popular at the turn of the twentieth century. Criollo were brought to the Southwest more than four hundred years ago by the Spanish and are well adapted to arid lands.

"My interest in these animals began while reading a ranching magazine published early in the last century," said Ed. "In this publication, a rancher was concerned about bringing Hereford cattle into New Mexico because the little Spanish cattle could "rustle up grub" better than any other cow he'd seen. In a world of increasing energy costs, and human competition for grains, an animal that can rustle up its own grub might be just what the industry needs."

To find out, Ed and Mexican researchers selected semiwild Criollo cattle from the Copper Canyon country of northern Mexico and brought them to the JER. Their goal is to understand what behavioral and physiological traits allow Criollo to persist in arid environments. For instance, does their relative "wildness" give them an adaptive advantage when it comes to disease? He also wonders what role Criollo cattle might play in the development of alternative livestock production systems that could fit desert environments. Could they be good grass-fed animals that might become a part of the burgeoning health food market?

Ed's goal is to help discover the most resilient ranching for the region. "Prescriptive, or rigid, production systems lead to greater dependency on others, and ultimately fragile systems," he suggested. "Knowledge-based systems lead to creativity and a greater ability to adapt to change. This is important. Although the beef cattle industry in the United States is highly efficient, it also has become increasingly centralized and rigid. This leads to a system that is vulnerable to collapse in response to catastrophic events. By promoting increased entrepreneurship, the system will become more diverse and increasingly modular with time; thus, it will be more resilient."

In a sense, the circle is closed: resiliency is also the key to ecological integrity, which is the foundation for economic sustainability.

Being useful in range science also means not sitting still for very long. In the case of the JER, this means tackling new problems, such as understanding how ecological processes work at landscape scales, how to effectively conduct rangeland restoration, what the ecological effects are of land fragmentation, and how best to mesh ecological flexibility with economic flexibility.

To accomplish these goals the scientists at the JER are taking an increasingly collaborationist approach. Under Kris's direction, all the scientists meet regularly to share their latest research as it pertains to overarching issues, such as developing land health indicators that are useful at landscape scales.

"You work for the good of the group," said Jeff. "To the extent possible, egos are jettisoned. Collaboration is the new paradigm." The researchers at the JER also understand, of course, that collaboration is also the key to getting anything done on the ground at landscape scales in the West.

"Fifteen years ago the question to us was: tell me how rangelands work," said Kris. "Today, the question we get asked is: how do we restore and maintain these systems?"

"What we've learned is that the trouble has not been with the tools," continued Kris, "but how we have used them without a landscape ecological perspective. But that's changing. There's a movement now—it's not just scientists. Politicians are talking about it, so are business leaders. It's going global too. Range health manuals are being translated into Mandarin and Mongolian as we speak."

It will be difficult to accomplish landscape-scale restoration, everyone knows, because of the challenges posed by the long tradition of managing the West by "fractions," as Kris put it, referring to the fragmentation of lands among various private, state, tribal, and federal owners.

But this too is changing as tools such as grassbanks, for example, encourage management to be coordinated over larger landscapes. And there is little doubt that the JER will be right in the thick of this change as well, usefully teaching us that land health is the foundation to social and economic health.

Or, as Kris Havstad put it: "To be truly sustainable, we must be educated and practiced observers of our environment."

Chapter Thirteen

CEASELESS CHANGE

Utah State University *Logan, Utah*

What do the *Far Side* cartoons of Gary Larson, plant toxicity levels, the Buddha, intellectual paradoxes, Forrest Gump, and a brief history of western science have in common?

The short answer is that they are all part of a typical lecture by Dr. Fred Provenza, a professor in the College of Natural Resources at Utah State University.

The long answer is that they are integral parts of his thoughtful analysis of the interconnections among the habits of animals, the nature of scientific inquiry, the role of stewardship in natural resource management, and the knowledge that we live in a universe of ceaseless and relentless change. Provenza's analysis is invariably delivered with an infectious smile and a refreshing sense of humor, which is where the cartoons come in.

Forrest Gump, however, has a more serious job. In a lecture that I attended, Fred opened with what he considered to be Gump's dilemma: is life all about one's destiny, as the film's character of Lieutenant Dan believed—to die on the battlefield—or is it dominated by randomness, symbolized by a box of chocolates, as Forrest's mother suggested?

In an attempt to answer this meaty question, Fred followed with a lesson in the history of science.

For more than two hundred years, Newtonian physics taught three basic truths: (i) nature was knowable and predictable (i.e., mechanistic); (ii) a whole equaled the sum of its parts (i.e., reductionistic); and (iii) time and space were absolute and separate. The role of the scientist in this worldview was to discover the rules that governed how nature worked.

Then Einstein turned things upside down in 1905 by showing that time and space were relative—subject to forces such as gravity—as well as interconnected. This unsettling insight was followed closely by the even more disturbing development of quantum physics, which declared that although nature is knowable it is *not* predictable, and the sum of the parts does *not* equal the whole. In other words, as Fred put it, contrary to Einstein's famous declaration, God *does* play dice with the universe.

"Our Western culture teaches us to think in linear, hierarchical ways," he said. "In fact, there is no one central controlling force, only a large number of agents, all interacting and adapting to each other and to their local environments. Ultimately, a highly complex order emerges from the local interactions of all of the parts."

Life, in other words, is an ongoing series of choices made in the face of uncertainty and change. This idea has important implications for conservation, Fred argues, because just as physicists have been forced to relinquish their rigid Newtonian views, ecologists, environmentalists, animal scientists, and managers of natural resources need to abandon inflexible perspectives for ones that reflect the dynamics of life.

To make the point, he quoted the British novelist G. K. Chesterton: "All conservation is based upon the idea that if you leave things alone you leave them as they are. But you do not. If you leave a thing alone you leave it to a torrent of change."

All of which led Fred to frame a larger question, which pertains as much to individual human life as to animal life on a ranch:

> How does one manage ongoing interrelationships among facets of complex, wholly interconnected, poorly understood, ever changing ecological, cultural, and economic systems in light of a future not known and not necessarily predictable, in ways that will not diminish options for future generations?

In many ways, this question is the culmination of Fred's academic career, which has focused on animal behavior, livestock specifically and wildlife more generally. An interest in why animals do what they do developed early in his life. Growing up in the Salida Valley of southern Colorado and working on area ranches, he became fascinated by what sheep, cattle, deer, and elk ate on the open range and became curious about why. Observing that how people make a living in ranching often ignored "how animals make their living," Fred pursued his interest in range and wildlife science and animal behavior through a master's degree and a Ph.D. at Utah State.

What Fred learned is that we are all creatures of habit with often good reasons behind it, though sometimes it may not seem like it. Take nutrition for example. Herbivores eat a diverse array of species—as many as one hundred species—but studies showed that the bulk of a meal normally contains fewer than ten species, and typically as few as three to five. Their selections are guided by nutrients and toxins in foods, and they begin to learn these behaviors early in life from their mothers.

These studies demonstrated that herbivores are "nutritionally wise," which, according to Fred, contradicted the long-standing belief that herbivores are generally "unwise" because they don't always choose the most nutritious foods.

This belief created a paradox: we are often baffled when animal performance declines despite an abundance of suitable habitats and nutritious forage. One key to resolving this paradox, Fred learned, has been a clearer understanding of the role experiences early in life play in shaping diet and habitat selection behavior—essentially creating locally adapted animals that when moved to unfamiliar environments don't perform well.

Another key has been understanding the role toxins play in animal diet and the regulation they place on food intake—a role that influences behavior. By setting limits on intake of any one food, toxins force animals to eat a variety of foods to meet their nutrient needs. Moreover, every individual is different in its needs for nutrients and ability to cope with toxins. Thus, grazing practices that allow the individuality of animals to be expressed are likely to improve performance of the herd.

Another insight illuminated how animals learn. Fred and his

colleagues discovered that when young herbivores are encouraged to eat only the most preferred plants, they are not likely to learn to mix foods high in nutrients with foods that contain toxins. Experienced animals learn to eat a variety of foods, even when more nutritious foods are available.

Other insights that Fred and fellow researchers gained include the following:

- Because life for herbivores exists at the boundary between order and chaos, animals, humans included, learn habits to create order and predictability.
- The origins of animal food habits and habitat preferences involve interactions between the social organization (culture) of the herd and the individual.
- Although both people and herbivores strive for order, they also seek variety.
- Ongoing changes in social and physical environments require old dogs to learn new tricks all the time.

"Thus, while the behavior of herbivores may appear to be little more than the idle wanderings of animals in search of food and a place to rest," Fred summarized, "foraging is a process that provides insights into an age-old dilemma faced by herbivores and humans alike: how do creatures of habit survive in a world whose only habit is change?"

In other words, if we are, as Aristotle once remarked, "what we repeatedly do," then how do we break bad habits and manage for long-term sustainability?

The first step, according to Fred, is to try to understand what part of our behavior is cultural and what is not. Take, for instance, livestock grazing in riparian zones. Cattle are not genetically preprogrammed to wallow in riparian areas, said Fred. Instead, it is a learned behavior—a habit. "Cattle can be trained to prefer uplands over riparian areas, but only if people manage using behavioral principles," said Fred. "No gene codes for living in riparian areas."

Fred often cites the example of rancher Ray Bannister, who manages cattle on his property in eastern Montana according to "boom-bust" principles that require intensive, soil- and plant-stressing periods of heavy grazing followed by two years of complete rest. This system forces the cattle to eat all the forage in a pasture, not just the "ice cream" plants, thus eliminating the competitive advantage of the unpalatable plants.

As a result, said Fred, "It is hard to find any part of the ranch that lacks abundant plant cover even during years of drought."

Bannister's challenge, however, was convincing his Herefords to change their eating habits. It took three years—a typical period—for his animals to adjust, during which their weight and performance dropped dramatically before recovering. Now the mother cows teach their calves the system and all is well on the Bannister ranch.

In fact, Fred believes that management-intensive systems, like the one employed by Ray Bannister, balance animal, plant, social, and economic concerns. "These endeavors have made vast strides to integrate the science of plant-herbivore interactions with the art of grazing management across landscapes."

In Fred's view, we rely too much on technology and not enough on the culture inherent to social animals, in particular the collective knowledge and habits acquired and passed from generation to generation about how to survive in a particular environment.

"Unfortunately, social organization and culture are rarely considered important in the structuring and functioning of ecosystems," said Fred, "and indeed we manage wild and domestic animals in ways that thwart the development of cultures, perhaps to our long-term detriment."

If we instead allow cultures to evolve, we may lessen our dependency on technological fixes and come to rely more on behavioral solutions that cost very little to implement and are easily transferred from one situation to the next. Unfortunately, said Fred, scientists and managers often ignore the power of behavior to transform systems, despite compelling evidence. "Once mastered," he continued, "behavioral principles and practices provide an array of solutions to the problems people face in managing to improve the integrity of the land and to make a living from the land."

This is, of course, easier said than done because change often involves stress.

"The Buddha teaches us that all suffering arises from trying to cling to fixed forms—objects, people, ideas," said Fred. "The challenge is to accept the world as it moves and changes."

But it is also more than that, as Fred discovered in 1999, when he was diagnosed with colon cancer. The experience became a turning point for him. Acceptance wasn't sufficient. He had to, in his words, "get going," which included finding a way to make his research more accessible to those who might benefit from it.

In the following year, with a grant from the U.S. Department of Agriculture, Fred launched a major educational program called BEHAVE (Behavioral Education for Human, Animal, Vegetation and Ecosystem Management) whose mission is to inspire people to master and apply behavioral principles in managing ecosystems. His target was anyone managing land, and over the intervening years, Fred and his graduate students have taught BEHAVE classes all over the West.

His hope is for people of disparate beliefs to work together and thus move from observer to participator—to "get into the game," as he puts it, even if that means venturing into the unknown.

The familiar, the comforting, the orderly, and the predictable are devoid of creative zeal, he said. The unfamiliar, the obscure, the potentially dangerous, the unpredictable are pregnant with creative opportunities. Still, it is in the courage to love, Fred insists, that the hope of the world resides.

"The best way to predict the future is to create it, and in the arena of constant transformation, anything is possible if we dare to engage one another and the environment creatively. Ultimately, the courage to love is the courage to transcend boundaries and traditions, and it is the source of creativity."

SEEING THE FOREST
AND THE TREES

Bandelier National Monument *Near Los Alamos, New Mexico*

Dr. Craig Allen loves what he does.

This passion is not only evident in the many articles he has written for scientific journals, the many lectures he has given on forests and fire to a wide variety of audiences, and the elegant experiments in ecological restoration he is conducting, but also in the energy he puts into a conversation about forests while just *hiking* to a project site.

What Craig does is try to understand fundamental ecological processes in the woodlands and forests of the Jemez Mountains, west of Santa Fe, New Mexico. Employed by the U.S. Geological Survey and stationed at Bandelier National Monument, Craig has devoted more than twenty years of his professional life to gaining a comprehensive understanding of forest health, forest sickness, and what constitute appropriate cures.

But it is not merely an academic interest. Craig believes humans have a responsibility to repair damaged land and has become a vocal advocate for science-based "adaptive management"—carefully monitored experimentation to discover optimal conservation practices—in our forests.

He is keenly aware of the need to restore our forests to health. That's why he tries to make clear the relevance of his work to nonscientists.

"This is a harsh environment," Craig said during a hike to a restoration study site on a mesa in Bandelier's wilderness. "There's pounding rain in the summer, when it rains, lots of freeze-thaw action in the soil in the winter, when it snows, and desiccating winds in the spring, when it blows."

Combined with intensive human use of the Jemez Mountains—logging and grazing in particular—these conditions played havoc on Bandelier's delicate and shallow soils, which are not untypical of many around the state. And much of the state suffered a human history similar to Bandelier's: decades of rough treatment by settlers and others following the arrival of the railroad in northern New Mexico in 1880. The question is: what should be done now?

When we reach the restoration site, Craig explains what they did.

"In 1997, crews came in here and cut the trees, lopped the branches and spread everything out over the land," he said. "The idea was to get a more natural water cycle going by allowing more infiltration by rain so grass would grow. We wanted to do this by improving microenvironments in the bare interspaces between grass clumps and trees, and we did that with the slash." (Slash is the twigs and branches removed from the main trunk of a tree.)

"There was an immediate response," said Craig. "Remnant grass bunches started growing again, a weedy successional cycle started, and new plants grew."

The slash did this, according to Craig, for three reasons:

- The branches and needles increased "surface roughness" by creating a "zillion" little check dams that held back water.
- The decaying foliage provided a pulse of nutrients to plants and seeds.
- Shading by the branches reduced evaporation.

"By reducing the harshness of the microenvironment," he said, "we increased the amount of plant-available water, which is essential to slowing and stopping sheet erosion."

If rain runs off too quickly, plants can't grow, and if plants can't grow, they can't become fuel for a fire, and if a fire can't run its course, then too many trees grow, which reduces the amount of plant-available water, which encourages additional erosion, and round and round it goes, as it has for nearly a century.

When Craig and other scientists compared the restoration site to an adjacent control watershed that did not receive treatment, they were pleasantly surprised by the results. "Overall biomass went up six- to eightfold," Craig said, "and sediment yield dropped one hundredfold. Biodiversity and abundance went up too. We even started to see butterflies again because the plants were flowering."

"It was very encouraging," he continued. "It showed us that you can kick-start natural processes again without too much work or money. We didn't plant any seeds. All we really needed were chainsaws."

All of this represents a new approach to restoration. First and foremost, it's humble. "We can't erase history," Craig noted, "but what we can do is allow ecological processes to function again as naturally as possible. And be ready to admit mistakes."

Craig is the first to acknowledge that they don't know exactly when they will reach the endpoint of this experiment, but he does know they can't be managing it forever. "We don't want to be endlessly deciding who lives and who dies out here," he said. Their goal is to let nature take over as soon as possible and get fire back in the system.

Their approach is also practical. Craig thinks this method will appeal to landowners because of its simplicity. On a larger scale, with larger trees, he recommends the employment of a "splatterer"—a machine on rubber tires that "eats" trees from the top down using a fast-spinning rotary head and a rotating cab. Debris from this process is "splattered" for two hundred yards in a random manner that Craig considers to be natural enough.

He frowns on chipping, a popular process for wood removal that involves grinding small-diameter branches into small bits and then spreading the mulch over the ground. Craig believes it doesn't mimic natural processes properly.

"Popping trees out of the ground may not be enough either," he warned. "You're just reducing tree competition, not addressing the problem of poor water cycling."

Craig candidly admits that their approach may not be ideal for everyone, but some sort of approach is urgently needed. "We've got 100,000-year-old soils in Bandelier that will be gone in two centuries if we don't do something," he said. "Some might argue that we should sit back and let nature take its course, but I'm not one of them."

Thus, Craig joins the growing ranks of researchers, land managers, and others who see an important role for people *in* nature, principally in the role of restorationist. Whether it is forests or range or riparian areas, the job is principally the same: help nature get back into shape.

"Aldo Leopold observed years ago that many southwestern ecosystems were in trouble," said Craig. "They're still in trouble. The difference is today we now know enough to make progress in repairing the damage. We don't know everything, not by a long shot, but we know enough to get started."

Chapter Fourteen

MUGIDO

Teel's Marsh *Near Mina, West-Central Nevada*

Not long ago, I attended a meeting at the headquarters of the Bureau of Land Management (BLM) in Nevada where two ranchers, a husband and wife team, tried to convince the BLM to let them implement a visionary and audacious plan to restore life to Teel's Marsh. It was once a thriving terminal lake but it's now a lifeless salt flat. The ranchers argued passionately that they could revive the marsh by repairing the dysfunctional water cycle in the one-hundred-thousand-acre watershed. Their daring idea? Break up the capped soil (hardened in a way to make it impervious to water infiltration) with the ground-disturbing impact of a thousand, or more, cattle hooves.

The ranchers' credibility rested on their long experience in range restoration, including their success in creating life on sterile mine tailings through the "poop-and-stomp" action of animal impact. And their work was backed up by monitoring data, they explained.

They were supported at the meeting by a prominent environmental activist who had built a formidable reputation as an outspoken critic of the livestock industry. These ranchers were different, she insisted. She knew them to be careful stewards, having watched bird populations rise steadily on their grazing allotment for nearly a decade. And as a dedicated birder, she knew that the marsh was part of an important historical flyway in the region.

The BLM's response to their entreaties, however, was "no." The reasons cited were technical and bureaucratic: the grazing permit wasn't in order, old paperwork had been misfiled, the proper bureaucratic procedure had to be followed, archaeological clearance would have to be done, workloads were too heavy, staffing levels were too light, budgets were declining, demands were rising, and, ultimately, the admission that "higher ups" were too skeptical.

The ranchers responded by saying they would assume all the risk, including the financial cost, and do all the work. All they needed was a green light from the government. Teel's Marsh, part of a congressionally designated wild burro refuge (though overgrazed by burros, they noted), was essentially dead. It had nowhere to go but up, they said. They could do it.

"It'll never happen," said a sympathetic BLM range conservationist.

By the end of the meeting I was as frustrated and upset as the ranchers and the activist. That's because this is still a too-common story across public lands in the West today. Progressive, innovative proposals to repair damaged land, to employ new land management models, to implement "out-of-the-box" tools and ideas that produce results, too often meet the same fate: "No."

Unhappy thoughts about federal lands management is a relatively new and uncomfortable feeling for me. For all of my adult life, plus a few of my teenage years, I believed in the primacy of public lands. When traveling to national parks, for example, I invariably expressed this belief by writing in every visitor book: "Buy more land!"

I meant it too. Like many of my fellow urbanites, I believed the simple answer to the complex questions surrounding land use in the West was increased federal ownership, especially if it meant an expansion of our national parks.

My belief took root in my youth. Shortly before my sixth birthday, my family and I emigrated from Philadelphia to Phoenix in a covered station wagon, becoming part of the great demographic shift that would irresistibly transform another sleepy western town into a bustling, and apparently boundless, megalopolis.

My parents, like so many of their generation, had farm roots, though neither was interested in agriculture. As an unconscious compromise, perhaps, they moved us to what was then the edge of town and bought horses. This meant that I lived in two worlds as I grew. Driving through an asphalt jungle by weekday and prowling the desert on foot and horse-back by weekend, I careened back and forth between urban and rural, which, like so many of *my* generation, meant having the best of both worlds for a time.

A lot of this changed in a rush during the summer of my sixteenth year, when I took a backpacking tour of western national parks with high school chums. What I discovered, of course, was the federal commons. I learned that an invisible line separated public from private, wild from nonwild, commercial from noncommercial, sublime from soiled. I returned from this voyage of discovery believing wholeheartedly in the observation of Lord Bryce, who wrote years ago that our national parks were the "best idea America ever had."

Over the years, as my interest and knowledge about the American West grew, my core belief in the dominance of the federal commons remained unshakeable. It even survived my stint as an employee of the idolized National Park Service, where my exposure to the dysfunctional side of bureaucracies failed to rattle my faith in the preservation para-digm. If the federal government had warts, it was still preferable to any alternative.

I don't believe that anymore.

I still believe in the federal commons—the system of national parks, refuges, rangelands, and forests that comprises half of the land in the West. And I still support public lands for the same reasons I did as a youth: the democratic ideal they represent, the beauty and biodiversity they protect, and the bulwark against residential development they provide.

I am also aware of history—that the idea of public lands retention was forged on the anvil of hard use; that a late nineteenth and early twentieth century legacy of deforestation, overgrazing, and other forms of short-term exploitation of land and people contributed significantly to the popular demand for protection. And as long as the threat of hard use still exists—as unfortunately it does—the federal commons remains necessary.

But although the ideal is still valuable, its implementation has become a dilemma. Though it wrenches to say so, I'll put it bluntly: the old model of governance of these special lands is worn out. I believe this for the same reason that I think the traditional ranching and environmental paradigms are wearing out as well: old thinking and old structures have become obstacles to needed innovation.

The management of federal lands, proactive and innovative in the early years, has become today, for a variety of reasons, mostly about "no" (there are painful exceptions, such as the massive expansion of oil and gas development on federal lands currently underway). This is a dilemma, because although in recent years new ideas, new practices, new paradigms, and new values, as well as new threats, have emerged in the West, few of them can get past the "no" logjam on public lands.

Rather than despair, however, I began to look for a new model of public lands management that would serve as a starting point for a discussion on how to substantially reinvigorate what is still one of the best ideas we ever had.

A few years ago, the state of Colorado used lottery money to purchase a medium-sized ranch not far from a major city along the Front Range. The goal of the purchase was to protect open space in a rapidly fragmenting landscape, as well as to ensure environmental values for the long run.

The trouble was that the state had neither the capacity nor the desire to manage the land. So it issued a request for proposals (RFP) to see who might be interested in leasing the ranch. This was a competitive process, and, in fact, when the smoke cleared a rancher and his family had won.

The rancher promised to do the following:

- Make an annual lease payment to the state of Colorado
- Keep the land in agriculture

- Meet or exceed high environmental standards (documented by monitoring)
- Provide educational and other forms of outreach programs on the ranch, aimed particularly at the residents of the major city nearby
- Provide hunting and recreational opportunities to the public

In doing so he would accomplish the state's goals: open space would be protected and environmental values would be ensured.

In turn, the rancher received assurances from the state that he would be able to run the ranch as he saw fit, with a minimum of regulation. Most importantly, he would be allowed to make a profit (which enables him to make his lease payment). Regulation by the state was swapped for innovation, flexibility, and entrepreneurial energy on the part of the rancher. Colorado owned the land and retained oversight, including, potentially, enforcement of environmental standards, but otherwise it basically got out of the rancher's way.

Why can't a similar deal take place on federal land?

Many of us thought something of this nature might happen when the U.S. government purchased a ninety-eight-thousand-acre working cattle ranch, located in a large collapsed volcano above Los Alamos, New Mexico, and created the Valles Caldera National Preserve (VCNP) in 2000. The deal was brokered by New Mexico's senior senator, Pete Domenici, whose support was contingent on the creation of a new model of federal lands management. Apparently as frustrated with the logjam on the federal commons as anyone else, Domenici insisted that the VCNP be governed by a nine-member board of trustees, each representing a different "use" (wildlife, grazing, forests) of the land.

Although the legislative mandate of the board is to protect the conservation values of the property, Domenici also insisted that the board manage the preserve for eventual financial self-sustainability, truly a remarkable goal for public lands. The only other example in the nation of a board of trustees managing federal land for conservation and financial gain simultaneously is the Presidio, an old military fort located in the heart of San Francisco—a wholly different kettle of fish.

Today, the VCNP is still nowhere near financial self-sustainability;

and many observers, including this one, are doubtful that it will be able to achieve this important goal. Part of the trouble may be with the trust model—perhaps managing land by committee is easier said than done—or perhaps the trouble simply was elevated expectations. In either case, the VCNP "experiment" is beginning to look like a golden opportunity missed.

Take the livestock grazing program at the VCNP, for example. It has struggled from the get-go as a result of shifting directions from the board, unimaginative performance on the ground, and poor public relations. Worse, it has lost money every year of operation—an astonishing fact given that the grasslands of the preserve are some of the most productive in the Southwest.

Could things have been different? Instead of micromanaging the livestock program, could the board have done what the state of Colorado did: issue an RFP? Why not turn the grazing program over to a progressive land manager and let him or her do the work? If it were a matter of targets and conditions, such as environmental health, or educational activities, or outreach to local communities, why not write those conditions into the RFP? The roles of the board would have been then to provide clear objectives, do the monitoring, and collect an annual payment. (In late 2006, the VCNP issued just such an RFP, and in 2007 the winning rancher wrote a check to the Board of Trustees.)

I firmly believe that the grazing program on the Valles Caldera, in the hands of any number of progressive ranchers I know, could be ecologically robust, responsive to social and cultural needs, and economically profitable—profitable to the board (and thus the American people) as well as the rancher. And it could do so while being public land, owned and shared by all Americans.

I further believe this could be true of much of the federal commons. The rise of the progressive ranching model, coupled with an explosion of ecological knowledge and new methods of scientific documentation in recent years, means that there is no longer an intrinsic contradiction between commercial activity and ecological function on the range. This may have been the case once upon a time, but it is not now. The trouble, then, is not with the toolbox, or the profit motive. The trouble is with the model.

The examples of the Colorado rancher and the VCNP, coupled with my brief, but sobering, experience managing the Rowe Mesa Grassbank, a thirty-six-thousand-acre ranch on Forest Service land, have led me to a new idea for the federal commons.

I'll call it a "mugido"—the Spanish word for the moo or lowing of a cow (sounds like "ejido," which is the Spanish word for "commons")—though it can also be referred to as an "RFP" model.

A mugido is a stretch of public land where the government reduces its regulatory role in exchange for high environmental stewardship by a nongovermental entity. In a mugido the government's role is to set ecological and social standards and objectives through collaborative goal setting, provide technical assistance (fire, archaeology, biology), and conduct oversight and monitoring. The role of the private entity is to meet, or exceed, the collaboratively derived goals and objectives.

In other words, a mugido is an equitable public-private partnership. The land would remain part of the federal commons, still influenced by national and regional goals, still owned by the American public, but operated by a private entity in collaboration with the overseeing federal land agency.

For example, although the Forest Service or the BLM would set environmental goals for the allotment (or landscape), it would be the permittee's decision how to achieve them. The goals would be set collaboratively, drawing on each member's strengths, but the permittee would have discretion over the toolbox: what type of livestock to use, for example, their numbers, timing, and intensity.

The permittee would be empowered to be as innovative, flexible, and entrepreneurial as he or she wanted to be; and the government would retain the right to judge the effects of these actions and respond appropriately.

Not all regulation would disappear. Ensuring the recovery and maintenance of endangered plants and animals, for instance, would be subject to enforcement. But collaborative decision making coupled with innovative implementation of best management practices, audited by the government, means that the "hammer" of regulation could be laid down.

I need to be clear that by proposing a mugido model I am not trying to poke federal employees in the eye. Nearly all civil servants that I have met over the years are hard working, dedicated, and imaginative people. It's not their fault that the system has basically ground to a halt all around them. Rather, a mugido acknowledges their plight—declining budgets, increased workload, more and more layers of rules and regulations—and seeks to find a positive role, as partners, for them on the land.

Nor am I proposing that all federal lands become mugidos—far from it. In the beginning, in fact, mugidos will be few and far between. That's because they should be carefully created on a case-by-case basis and only when an allotment or permit has become "open"—i.e., when it has been vacated by its previous owner.

Another option would be to create a mugido when a current permittee is ready, willing, and able to make the transition. In either case, to succeed the private entity has to have the right set of skills, credibility, and financial resources to do the mugido right. At the same time, a mugido cannot be imposed by the government; it needs to be voluntary. And if a particular mugido doesn't work out, then the government reserves the right to go back to the old model.

The goal of a mugido is to get innovation on the ground by blending the best of both worlds: the entrepreneurial spirit of the private community (which includes nonprofits) and the big picture ideals of the federal commons. In other words, a mugido is all about "yes."

But what if the RFP results in an out-of-state entity taking away an opportunity from the local community, especially if that community is historically, socially, or economically disadvantaged?

I don't have a simple answer to this problem. Currently, grazing permits (and the private land they are attached to) can be bought and sold without regard to the needs of local communities. Ideally, mugidos would be locally based and would engage local communities. Perhaps this can be written into the RFP in some way—that local partnerships are paramount or that the mugido must serve local interests to a significant degree.

Balancing local, regional, and national needs will be a central task of a mugido.

Obviously, this is a controversial idea, and undoubtedly there will be objections. But let me try to sort out what I see as the five key elements of the mugido model:

1. *The overarching goal is land health.* The basic idea behind land health is that by restoring and maintaining land *function*—what Aldo Leopold called the "land mechanism"—we can create a solid foundation for the social *values* we place on the land. In other words, if we jeopardize or degrade function (soil stability, water and nutrient cycling), then the land's ability to support our values (food, water, wildlife, recreation, grazing) will eventually degrade too.

Jared Diamond's book *Collapse: How Societies Choose to Succeed or Fail* documents in sobering detail what happens to communities and cultures when land function fails. Fortunately, advances in ecological knowledge, coupled with new quantitative and qualitative monitoring protocols, have made it possible to develop a much clearer picture of what land health means than we had sixty years ago when Aldo Leopold coined the term. This means that land health targets can be described, measured, and analyzed. They can be achieved too, as well as enforced if necessary.

On public and private land, the bottom line is land function, from the soil up. If land exists in a degraded condition and is in need of restoration, then that should be the primary goal of its managers. If it is healthy, then it needs to be maintained. Unfortunately, much of the West is degraded, for a variety of reasons, including much of the federal commons. Tackling this land health crisis, principally through restoration, will require a great deal of innovation, education, and commercial activity.

2. *The whole toolbox is available.* Achieving and maintaining land health requires having the entire toolbox (rest, grazing, fire, thinning, restoration, etc.) at one's disposal. It also requires having the flexibility and incentive to quickly choose a particular tool for a particular job. Nature

is not static—it exists in a constant state of flux, including sometimes violent perturbations. Stewardship, especially restoration work, needs to be equally active, within the limits set by collective goal setting. Evaluation of the effectiveness of any particular tool is necessary as well.

But the freedom to innovate is necessary too. The power of creativity needs to be tapped, encouraged, and rewarded, especially given the scale of the task of stewarding land today. The initial response by the government to a new idea should be "why not?" If implemented, it should then be followed by monitoring, evaluation, and adjustment. Regulation should follow innovation at a distance, not stand in its way.

In a mugido, the principal role of the government is that of an auditor. It should check progress one, two times a year, maybe more, and suggest, or require, changes if necessary. If a permittee has abused a tool, or failed to perform to predetermined standards, then the government reserves the right to terminate the relationship.

3. *Positive incentives (including profit) are emphasized.* The key to innovation is positive financial incentives for restoring and maintaining land health. Additionally, delivering values that society wants must result in a profit for the steward. Negative incentives—the threat of regulation, for instance, or paying a land manager *not* to damage land (the traditional response of government)—won't work in the long run.

But the answer doesn't lie wholly in the market either, not as long as it remains more profitable to exploit natural resources for short-term gain. Until we can create a "healing" economy—one that pays landowners and managers to restore and maintain land health on par with what they can earn by damaging land function—the marketplace cannot be allowed to have a completely free hand.

The answer, in the meantime, is to create a public-private partnership that is profitable to both, ecologically and economically. Private entities would be free to be entrepreneurial on public land, within limits enforced by monitoring, and public agencies would benefit from increased land health. Jobs would be available locally, which would help maintain community health. The best "yes" of all is a paycheck.

Right now, the incentives on public land all point in the wrong directions. Many grazing permittees feel little or no incentive to improve their stewardship, partly because they are not rewarded financially for it (and are sometimes punished) and partly because they consider stewardship to be "the government's job." That's often the problem with regulation—good stewardship needs to be encouraged and rewarded, not policed. And for federal employees there is little or no incentive to think outside the box. Too often, individual initiative hits a brick wall of bureaucratic indifference.

4. *Let government employees be free.* Most public land managers don't want to be regulators. They didn't go to college to study how to be bureaucrats. They studied natural resource management, or biology, or archaeology, or planning. They went to work for the government because they wanted to be foresters, range managers, biologists, archaeologists, and planners. They wanted to be outdoors, in the woods, on a horse, doing research, or setting a prescribed fire. They didn't go into government to enforce compliance, sit in a cubicle, push paper, or appear in court.

Government employees need to be professionals again. Let them get to "yes" by being biologists and archaeologists; let them monitor, and teach, and learn. Let them *help.*

Because private entities often won't have the technical or educational experience needed to understand all the variables of stewardship, this expertise can be provided by the government. The complex issues surrounding endangered species, for instance, require the involvement of people with specialized knowledge. This will be tricky because the intersection of wildlife management and land health, not to mention best management practices, is not yet fully articulated. But letting biologists be biologists, for example, is the first step.

This way they can become genuine partners in land stewardship.

5. *Find a role for urban folks.* The widening urban-rural divide is having deleterious effects across the West, politically, economically, culturally, and ecologically. As the West continues to urbanize at a rapid rate, and as city dwellers move to the country (or at least purchase big parts of it), the rift threatens to grow. Fortunately, efforts to close this divide are becoming more numerous, especially around organic farming, eco- and agrotourism, water quantity and quality issues, and the protection of open space.

An effort needs to be made to bridge the urban-rural divide on the federal commons as well. In particular, urbanites who care about the condition and fate of public lands need to be given an alternative to conflict. Right now, the principal way a city-bound person can express concern for a national forest or park is to write a check to a watchdog environmental organization, whose typical response to activities on public lands is often "no" as well, and often for good reason. There's always a bad dam, development, or oil or gas well to fight someplace. Fighting is as necessary as it is unfortunate.

What is needed now is a way for urbanites to say "yes" on public land. Restoration is one way: the physical process of getting out on the land and helping to heal a creek or a meadow with one's labor is a satisfying experience. Another is to become active in the stewardship of rural public land. Lend a hand, buy local food, invest in a cow, do monitoring, take a tour.

At the same time, permittees on the federal commons need to find positive roles for urbanites. Pull them in, get them involved, make allies. Take their money, and give them a return on their investment. Make them part of the solution.

Ultimately, a mugido is all about healthy relationships.

Let's go back to the meeting in the BLM headquarters in Nevada for a moment. The ranchers were proposing to restore Teel's Marsh to health through the innovative use of livestock. Their goal was to restore function to the one-hundred-thousand-acre watershed that surrounds the marsh by repairing the damaged water cycle, principally by breaking up capped soil so that water and seeds could do their thing.

They were proposing to carry the financial risk, as well as reap any financial reward. They also proposed to do all the work.

It's a radical and audacious idea, granted. But what if the BLM said "yes?"

What if BLM employees sat down with the ranchers and worked on a set of goals, including ecological benchmarks, for the watershed? What if they pledged to do the monitoring, as well as provide the archaeological clearances and other technical support the ranchers needed? What if they provided the oversight needed to satisfy various public values, such as recreation, in the watershed?

What if they then became partners in what happened next?

The ranchers and their collaborative team, which would include environmentalists, could then go to work. They would have the flexibility to improve the watershed with whatever tools they thought appropriate, under the goal-setting guidelines, whenever, and for however long they thought necessary.

They could find creative ways to engage urbanites in their project. Horse owners could herd cows; schoolchildren could monitor land health; the team could create a nonprofit organization called Friends of Teel's Marsh; urban elbow grease could be applied to the land.

In the meantime, the ranchers would be evaluated by the quality of their product: the restoration of the marsh. Hopefully, the evaluation won't be too harsh or hasty; restoration is slow business, especially in a desert. But periodic review by the government would serve as a reality check on the project. Were the ranchers moving toward their goals? Did the goals need to be revised? What worked? What failed?

Products would also include better communication, increased trust, stronger relationships, and true adaptive management. The marsh might even be restored! Or maybe not. Ultimately, the skeptics could be right. Maybe the marsh couldn't be restored. Maybe cattle are the wrong tool. But we will never know if we don't let them try.

When I visit national parks today with my family, including our nine-year-old twins, Sterling and Olivia, I no longer write "Buy More Land" in the guest ledger. If I have time to write anything I usually dash off a note of thanks—thanks to staff for doing a good job, thanks to the park for simply being there, or thanks to the idea of the federal commons in general. Like all good ideas, however, it needs an upgrade—it needs to catch up to the times. Clinging to the fixed form of the past has made the federal commons seem rather anachronistic, I suspect, to the next generation. We need to find a way to make our public lands new again.

For one thing, the twenty-first century—*Sterling and Olivia's century*—

is already shaping up to be a turbulent ride. The relentlessness of change may speed up in ways we can't imagine or anticipate; all we can say for certain is that things will look a lot different in fifty years. Hopefully, the next generation will not see the federal commons as anachronistic at all, because the democratic ideal it represents is not anachronistic, nor is the beauty it shields, the history it teaches, the wildlife it harbors, the opportunity to be in nature it provides, and the shared responsibility it represents. All of this, I'm convinced, will become more important, not less, as the tumultuous adventure of the next fifty years takes off.

In fact, I believe that nearly all of the lessons I learned over the past ten years—the need for more stewardship, not less; the willingness to learn, and to teach; the possibilities of giving, not just taking, and the role of *vision* in all that we do—will be of paramount importance. Ceaseless change means we must be ceaselessly adaptive. If we cling to fixed forms, we consign ourselves to endless stress and anxiety. And if some of us don't want to, or can't, change for whatever reason, I am comforted with the knowledge that our children will do the changing anyway, abetted by Mother Nature.

I hope that Sterling and Olivia, and their cohort, find getting into the game as exciting, hopeful, and inspiringly complex as I have. As Wendell Berry said at our annual conference in January 2007, of the challenges facing all of us in this new century: "We are not walking a prepared path." It will take many leaders to prepare this path. Fortunately, many of the stepping-stones we will need have already been discovered. It's up to us to get busy putting them in place so that the next generation may follow.

Acknowledgments

Although there is often only a single name on the cover, a book is a collaborative endeavor, the product of lengthy and intense dialogues with many wonderful people. That's especially true with this book. What ultimately came to reside between these covers represents a decade of travels, conversations, interviews, workshops, and teachable moments with others. The learning curve is steep, and I owe a great deal of gratitude to all my teachers and peers who took time away from their busy lives to help me understand these issues.

I'd like to start by thanking the individuals whose stories occupy the pages of this book for being both revolutionary and generous. Each is patient, kind, and inspiring. Three people in particular became mentors: Kirk Gadzia not only taught me the fundamentals of range and livestock management, but motivates me with his thoughtful, balanced, and humorful approach to life. Kris Havstad taught me not only the value of useful science, but also, in an age dominated by increasingly simplistic explanations, the importance of viewing the world in all its complexity. And Bill deBuys is a role model both as a leader of the "radical center" and as an eloquent articulator of good ideas. But everyone I met is a teacher of some sort, and I am indebted to each for sharing his or her knowledge and good deeds.

There are also many people who I met over the years, on many travels, whose stories did not make it into this book. They too are path makers and revolutionaries in their own right. The list runs long, and here I'll single out as many as possible: Sam Montoya, of Sandia Pueblo; Duke and Janet Phillips, of the Chico Basin Ranch; Bill McDonald and the Malpai Borderlands Group; Nathan Sayre, of the University of California, Berkeley; Linda Poole, of the Matador Ranch, as well as the members of the nearby Ranchers Stewardship Alliance; Dan Kemmis, former mayor of Missoula, Montana; Tony and Jerrie Tipton, of Mina, Nevada; Jim and Joy Williams, of Quemado, New Mexico; Gail Garber, of Hawks Aloft; Mac and John Donaldson, of the Empire Ranch; the ranchers of the Altar Valley Conservation Alliance; Sunny Hill, of Rainbow Ranch; Dave Ogilvie, of the U Bar Ranch; Diana and Alan Kessler, of Orme Ranch; Doc and Connie Hatfield; Jack and Pat Hagelstein; Joe Reddan, Dave Stewart, George Long, Judith Dyess, John Pierson, Al Medina, Ron Thibedeau, and Beverly deGruyter, all of the U.S. Forest Service; Steve Fischer, of the Bureau of Land Management; Susan Holtzman and the other members of the interagency National Riparian Team; Gary Nabhan, of Northern Arizona University; Mandy Metzger, of the Diablo Trust; Jan-Willem Jansens, of Earth Works; Curt Meine; Deborah Madison; Susan and Tom Simons; Dennis and

Trudy O'Toole; Cheryl Goodloe; Mary Riseley; Steve Carson, Van Clothier, Craig Sponholtz, and Steve Vrooman, restoration entrepreneurs all; Marty Peale, Doug Fraser, Norma McCallan, Don Goldman, and George Grossman, all peers in the environmental community; John and Joan Murphy, and Nan and Dick Walden, from Arizona; the University of Arizona Press; all the fabulous speakers at our annual conferences, and everyone else who has lent us a hand.

Many of these stories began as essays published in *Headwaters News,* an online newspaper (www.headwatersnews.org), and I'd like to thank Greg Lakes and Shellie Nelson, *Headwaters'* editors, for their support.

The hard work of The Quivira Coalition family made this book possible as well. Thanks first to Jim Winder for providing the inspiration to get The Quivira Coalition started and to Barbara Johnson for the pivotal role she played in helping the organization grow and succeed. Thanks next to Tamara Gadzia for her extraordinary skill, fortitude, and dedication; to Sheryl Russell, Craig Conley, Catherine Baca, Deborah Myrin, Michael Moon, Veronica Medwid, and Mike Archuleta, for *their* incredible skill, fortitude, and dedication. Also, Will Barnes, Rich Schrader, Kim McNulty, Mike and Pat Boring, Priscilla Stollenwerk, Tarry Pesola, Bill Corcoran, Avery Anderson, Mike Bain, Dawn Moon, Mike Belshaw, Larry Cary, and all our other wonderful friends.

The Quivira Coalition's board of directors has, and continues, to ably assist us: Mark McCollum, Frank Hayes, Dan Dagget, Jim Winder, Merle Lefkoff, Virgil Trujillo, Roger Bowe, Sterling Grogan, Bob Jenks, Ernie Atencio, Dutch Salmon, Kris Havstad, Ed Singleton, Joan Bybee, Maria Varela, Rick Knight, Tim Sullivan, Andy Dunigan, and Ray Powell.

And a special acknowledgement and thank you to Cullen Hallmark, who continues to provide us with indispensable advice and friendship.

Key too were the principal supporters of The Quivira Coalition, without whom we would not have accomplished half of what we have (including much of the travel documented in this book): Gene Thaw, Susan Herter, Sherry Thompson and the Thaw Charitable Trust; Owen Lopez and Norty Kalishman and the McCune Foundation; Art Ortenberg and the Claiborne-Ortenberg Foundation; Steve Rasmussen and the Messengers of the Healing Winds Foundation; Clint Josey and the Dixon Water Foundation; Jay Knight and the Knight-Bradshaw Foundation; Judith McBean and the McBean Foundation; Karen Griscom, Kimmie Green, and the New Cycle Foundation; Ed and Trudy Healy and the Healy Foundation; the Earth Friends Foundation; the Bybee Foundation; Tim Herfel, of the U.S. Environmental Protection Agency; Marcy Leavitt, Abe Franklin, Michael Coleman, Maryann McGraw, and others with the Surface Water Quality Bureau of the New Mexico Environment Department;

and David Chase, Margo Cutler, Joe and Valer Austin, Joan Bybee, Kathi Wunderlich, Jim Weaver, and many others.

Gene Thaw deserves special acknowledgement—he took a gamble and gave us our first grant and has steadfastly supported The Quivira Coalition over the years.

Jonathan Cobb, my editor at Island Press, gently reshaped the book into a much better product than what he originally received; Barbara Dean, also at Island Press, gave the project early support; and Rick Knight acted as my "agent," encouraging me to submit the manuscript. Without his initial push, I'm not sure what would have happened to this project.

I owe a special debt to Wendell Berry, a friend and a mentor as well, whose incredibly generous decision to publish my "Working Wilderness" essay in his collection of 2005 inspired me to pull this book together. If what goes around comes around, as I believe it does, then I hope I can be as inspirational to others as Wendell is to me.

Finally, to my family—Gen, Sterling, and Olivia (and Madeleine, our beloved dog)—who are the ultimate source of my inspiration. It is a joy and a privilege to travel with them, to learn with them (and learn from them), to share hopes and fears and memories, and to build the future together.

I can't wait to see what happens next.

Selected Bibliography

Tony and the Cows: A True Story from the Range Wars. By Will Baker. University of New Mexico Press, Albuquerque, NM, 2001.

Focusing on Tony Merten, a crusading antigrazing activist who took his own life, this brief but insightful book analyzes the "grazing wars" that dominated the 1990s. It concludes with commentary on environmental conflict.

Beyond the Rangeland Conflict: Toward a West That Works. By Dan Dagget with portraits by Jay Dusard. Good Stewards Project, Flagstaff, AZ, 1998.

This is the Pulitzer Prize–nominated book whose publication in 1995 marked the beginning of the end of the grazing wars in the American West.

Holistic Management: A New Framework for Decision Making. By Allan Savory, with Jody Butterfield. Island Press, Washington, D.C., 1999.

The first edition of this classic treatise on land management ignited a progressive ranching movement around the world. It is still a must-read for anyone concerned about sustainable use of our natural resources.

Working Wilderness: The Malpai Borderlands Group and the Future of the Western Range. By Nathan F. Sayre. Rio Nuevo Publishers, Tucson, AZ, 2005.

This book is an excellent introduction to the complex story surrounding livestock grazing in the Southwest. It tackles endangered species, range ecology, history, economics, and the "radical center" through the prism of an award-winning collaborative group.

Ranching West of the 100th Meridian: Culture, Ecology, and Economics. Edited by Richard L. Knight, Wendell C. Gilgert, and Ed Marston. Island Press, Washington, D.C., 2002.

Combining scientific analysis with eloquent poetry and prose, this collection of testimonials by people intimately involved in modern ranching gives a unique overview of working landscapes in the American West at the start of the twenty-first century.

The Art of the Commonplace: The Agrarian Essays of Wendell Berry. Edited and introduced by Norman Wirzba. Shoemaker & Hoard, Emeryville, CA, 2002.

This is essential Wendell Berry and essential reading for anyone trying to understand our modern predicament. Berry argues that our industrial economic model lies at the heart of our troubles, and that reinvention of an agrarian model is necessary for our survival.

Farming with the Wild: Enhancing Biodiversity on Farms and Ranches. By Daniel Imhoff and Roberto Carra. Sierra Club Books, San Francisco, 2003.

This book is filled with examples of eco-agriculture, whose practices have positive benefits for wildlife as well as the natural processes that sustain us.

The Sunflower Forest: Ecological Restoration and the New Communion with Nature. By William R. Jordan, III. University of California Press, Berkeley, CA, 2003.

This is a thoughtful meditation on the meaning of ecological restoration, authored by the former director of public outreach at the University of Wisconsin's arboretum. It offers a hopeful look at the future of a new conservation paradigm in America.

Second Nature: A Gardener's Education. By Michael Pollan. Grove Press, New York, 1991.

Using gardening as a touchstone, Pollan examines the complex and ever-changing relationship between humans and nature with humor and insight. It also provides a provocative review of the modern environmental movement.

This Sovereign Land: A New Vision for Governing the West. By Daniel Kemmis. Island Press, Washington, D.C., 2001.

Half of the American West is publicly owned, much of it managed by the federal government. This book posits a vision for these lands that emphasizes watershed-scale management, grassroots collaboration, and a higher level of accountability.

Index

About the Author

A former archaeologist and Sierra Club activist, Courtney White voluntarily dropped out of the "conflict industry" in 1997 to co-found and direct The Quivira Coalition, a nonprofit organization dedicated to building bridges between ranchers, conservationists, public land managers, scientists, and others.

His work with Quivira includes progressive ranch management, scientifically guided riparian and upland restoration, land health assessment and monitoring, workshops, outdoor classrooms, lectures, publications, site tours, capacity building, collaborative demonstration projects, a journal, and an annual conference. See www.quiviracoalition.org.

His writing has been published in various books and magazines, including *Headwaters News*, *New Farm*, *Rangelands*, and *Farming*.

He lives in Santa Fe, New Mexico, with his family and a backyard full of chickens.

About Island Press

Island Press is the only nonprofit organization in the
United States whose principal purpose is the publication
of books on environmental issues and natural resource
management. We provide solutions-oriented information
to professionals, public officials, business and community
leaders, and concerned citizens who are shaping responses
to environmental problems.

Since 1984, Island Press has been the leading provider
of timely and practical books that take a multidisciplinary
approach to critical environmental concerns. Our growing
list of titles reflects our commitment to bringing the best
of an expanding body of literature to the environmental
community throughout North America and the world.

Support for Island Press is provided by the Agua
Fund, The Geraldine R. Dodge Foundation, Doris
Duke Charitable Foundation, The Ford Foundation,
The William and Flora Hewlett Foundation, The Joyce
Foundation, Kendeda Sustainability Fund of the Tides
Foundation, The Forrest & Frances Lattner Foundation,
The Henry Luce Foundation, The John D. and Catherine
T. MacArthur Foundation, The Marisla Foundation, The
Andrew W. Mellon Foundation, Gordon and Betty Moore
Foundation, The Curtis and Edith Munson Foundation,
Oak Foundation, The Overbrook Foundation, The David
and Lucile Packard Foundation, Wallace Global Fund,
The Winslow Foundation, and other generous donors.

The opinions expressed in this book are those of the
author(s) and do not necessarily reflect the views of these
foundations.